MSRI Mathematical Circles Library

Experimental Mathematics

V. I. Arnold

Translated by Dmitry Fuchs and Mark Saul

Mathematical Sciences Research Institute
Berkeley, California

AMERICAN MATHEMATICAL SOCIETY
Providence, Rhode Island

Translated from Russian by Dmitry Fuchs and Mark Saul.

The Russian original was published by Издательство МЦНМО under the title Экспериментальное наблюдение математических фактов [Russian], © 2006. The present translation was made for MSRI, and is published by permission.

This volume is published with the generous support of the Simons Foundation.

2010 *Mathematics Subject Classification*. Primary 00A05; Secondary 34C07, 68Q17, 20B30, 11P99.

For additional information and updates on this book, visit
www.ams.org/bookpages/mcl-16

Page 159 constitutes an extension of this copyright page.

Contents

Preface to the English Translation

Vladimir Arnold was one of the great mathematical minds of the late 20th century. His work was of great significance to the development of many areas of the field. On another level, Russian mathematicians have a strong tradition of writing for, and even directly teaching, younger students interested in mathematics. This work is an example of Arnold's contributions to the genre.

In 2005, Arnold gave lectures at the Dubna summer camp. This camp is an extraordinary gathering of the Russian mathematical community, in which distinguished mathematicians work to support advanced high school and undergraduate students entering the field. The present book is based on notes from these lectures. As the reader will see, Arnold was very connected to the new generation of mathematicians. One can sense the urgency he felt at delivering his thoughts into hands that might take them farther. The reader expecting a formal mathematical exposition will sometimes not find it here.

One might mistake this style of the work as not just urgent, but sloppy. No. The style is well thought out. Arnold's approach to mathematics—and he makes this quite clear in several passages—was fluid and intuitive. He saw mathematics not as a flat plain to be surveyed, but as a rugged terrain to explore. The most exciting aspect of mathematics, for Arnold, seems to have been a dynamic search for pattern through examination of many special cases. That is, he held a severely Platonic view of the subject, as one that proceeds as if it were an experimental science—hence the title. After this exploratory phase, on can tuck in the ragged edges. Arnold does this in many—but not all—cases, giving us theorems and proofs in the classic manner.

But it is in the chase, in the experimental "phase" of the process of doing mathematics, that Arnold here seems to take the most joy, and offers this joy to a new generation. Mathematical mainstream culture, in which one burns one's scrap work, discourages this. Few mathematicians—indeed few scientists in any field—open their minds so completely as he has to their students.

Arnold's style is unforgiving. The reader—even the professional mathematician—will find paragraphs that require hours of thought to unscramble. In some cases, Arnold collapses an argument into a few sentences that might take up several pages in another style of exposition. In other cases, he gives an intuitive argument in place of a rigorous one, leaving the reader to construct the latter. He probably felt that the real work was done on the intuitive level, and that his teaching would be the more effective if he left the tidying up to the student. The reader must have patience with the ellipses of thought and the leaps of reason. They are all part of Arnold's intent.

These lecture notes were gathered in haste from the field, and we have corrected numerous misprints and small errors in notation. We have given several extensions—in Arnold's own style—to the work, in "editors' notes". At the same time, we have striven to deliver intact the style of the work. Arnold's mind leaps from peak to peak, connecting disparate areas of mathematics, all (or most) accessible to the student with an advanced high school education. And yet there is a unity to each lecture, a flow from very simple questions to deep intellectual inquiry, and sometimes right to the edge of our knowledge of mathematics.

We hope that we have preserved this coherence, but also the excitement of the work, the sharp, jagged edges and breathtaking jumps that characterize the author's thinking.

It is our pleasure to acknowledge the contributions of several colleagues to this work. In particular, Sergei Gelfand, at the American Mathematical Society, kept us on track at several key junctures. James Fennell sedulously proofread the manuscript and corrected the TEX files. We would also like to thank the students of the Gradus ad Parnassum math circles at the Courant Institute, who gave us feedback about several sections. Much of this work was supported by a generous grant from the Alfred P. Sloan Foundation.

MARK SAUL
February 2015

Introduction

> "Not having achieved what they desired, they
> pretended to desire what they had achieved."
>
> *M. Montel*

In this series of lectures I will talk about several new directions in mathematical research. All of these are based on the idea of numerical experimentation. After looking at examples such as $5 \cdot 5 = 25$ and $6 \cdot 6 = 36$, we advance an hypothesis, such as $7 \cdot 7 = 47$. Further experimentation either supports or disproves it.

For example, Fermat's hypothesis (that the equation $x^n + y^n = z^n$ cannot be solved in natural numbers with $n > 2$) was advanced as a result of his attempts at a solution. This hypothesis led to the creation of a whole field of knowledge, but it was proved only after a few hundred years had passed.

The majority of hypotheses that we make are so far not proven (nor refuted). I decided to give these lectures exactly because of my hope that the listeners will help in the investigation of these problems, if only by conducting numerical experiments (which I have also conducted, without a computer, in the bounded region of the first million numbers).

Lecture 1

The Statistics of Topology and Algebra

> "I never heard of such a mathematician: he is actually a physicist."
>
> *Landau, on Poincaré*

> "It is not Shakespeare that most matters, but commentaries on his work."
>
> *A. P. Chekhov, as described by B. L. Pasternak*

Poincaré, the greatest mathematician of the recent era, divided all problems into two classes: binary problems and interesting problems. Binary problems are problems which admit of an answer "yes" or "no" (for example, Fermat's question).

Interesting problems are those for which an answer of "yes" or "no" is insufficient. They require investigation of questions that lead one further. For example, Poincaré was interested in how to change the conditions of a problem (for instance, the boundary conditions of a differential equation), while retaining the existence and uniqueness of its solution, or how the number of solutions varies when we make some other change. Thus he started the theory of bifurcations.

Three years before Hilbert gave his list of problems, Poincaré formulated the basic, in his view, mathematical questions that the nineteenth century would leave for the twentieth. This was the formulation of the mathematical basis for quantum and relativistic physics.

Today, many people think that relativistic physics at the time, in 1897, did not yet exist, since Einstein published his theory of relativity only in 1905. But Poincaré formulated the principle of relativity earlier, in his article of 1895, "On the Measurement of Time", which Einstein actually used (and which, by the way, he didn't acknowledge in writing until 1945). In just the same way, Schrödinger, in laying the foundation for quantum mechanics, achieved his success only because he used the mathematical works of his predecessor Hermann Weyl, whom no one mentioned later on, although Schrödinger actually references these works (in his first book).

1. Hilbert's Sixteenth Problem

Although I basically agree with Poincaré, today I will talk about a binary problem (or almost binary: this is why I am going to talk about of it) posed by Hilbert, the 16th in his list.

This problem is actually much older than Hilbert's list. In general, it is one of the fundamental problems of all of mathematical science (and of many of its applications).

Here is a very simple example: for an algebraic polynomial f in two variables x and y, we look at the curve along which it equals zero:

$$\{(x, y) \in \mathbb{R}^2 : f(x, y) = 0\}$$

The problem consists in determining *the possible topological structures of this curve, if f is a polynomial of a given degree n.*

For example, if $n = 2$, then by the ancient theory of conic sections, the curve can be an ellipse, a hyperbola, a parabola, a pair of lines (which might possibly coincide), or the entire plane (if the polynomial is identically 0).

Augmenting the plane with points at infinity turns it into the projective plane, for which the problem becomes easier. (An ellipse, a hyperbola, and a parabola have the same structure in the projective plane. The only difference is in the position of this "circle" with respect to the line at infinity. See Figure 1.)

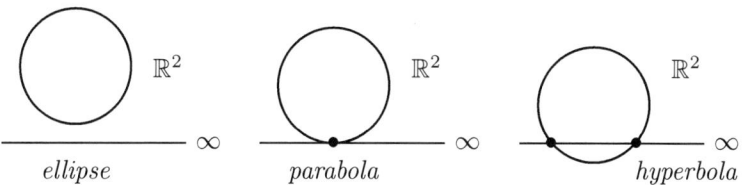

ellipse   parabola   hyperbola

Figure 1

For $n > 2$ the question is more difficult, but Descartes and Newton had already analyzed the cases $n = 3$ and $n = 4$. Hilbert asserted the he had looked into it for curves of degree $n = 6$, but his result (the proof of which he never published) was erroneous.

According to a theorem of Harnack, a curve of degree n consists of no more than $g + 1 = \dfrac{(n-1)(n-2)}{2} + 1$ connected components (where g is the genus of the associated Riemann surface formed by the complex solutions of the equation of the curve in the complex projective plane $\mathbb{C}P^2$). According to a theorem in topology, every closed connected orientable surface is a surface of genus g, where g is the number of handles we must affix to a sphere in order to obtain this surface (see Figure 2).

For $n = 6$ we find that the genus of the Riemann surface g is 10, so that a real curve of degree 6 has no more than 11 components (which are called "ovals", and resemble circles, or at least are diffeomorphic to the circle S^1).

Hilbert asserted that if the number of ovals is maximal (that is, if there are 11 ovals), then these 11 ovals can be placed on the (projective) plane $\mathbb{R}P^2$ *in only two ways.*

Each oval bounds a "disk", diffeomorphic to the interior of a circle. (The complement of this disk in $\mathbb{R}P^2$ forms a Möbius band: this is how Möbius discovered his surface.)

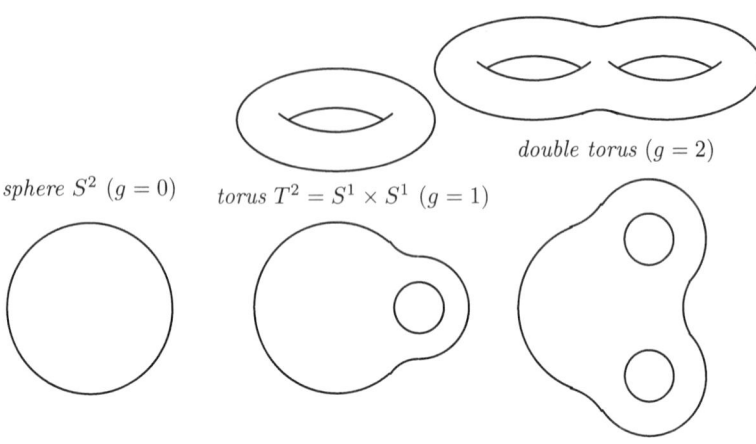

Figure 2. Surfaces of genus 0, genus 1, and genus 2.

And so, Hilbert asserted that only one of these disks can contain any other ovals inside it, and the number of interior ovals can only take on two values: 1 and 9 (Figure 3).

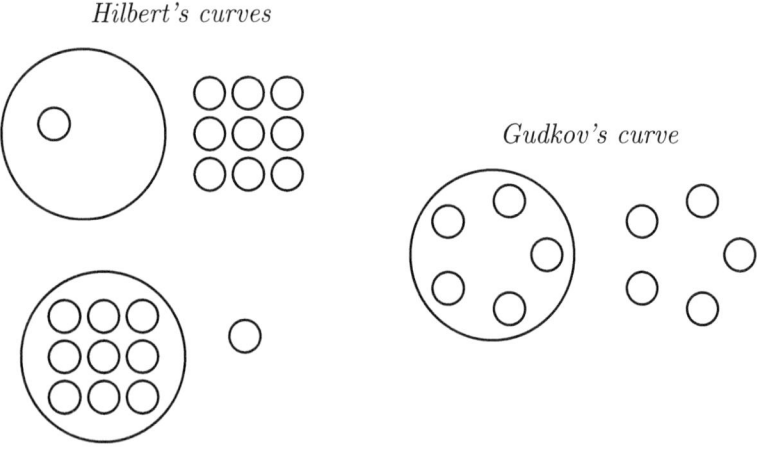

Figure 3. Algebraic curve of degree 6 with 11 ovals.

Hilbert's error consisted in the fact that the number of interior ovals could also be equal to 5. (This was discovered by Dmitri Andreevich Gudkov, a mathematician from Nizhny Novgorod, around 1970.) (See Editor's note 1, page 55.)

For curves of degree 8 Hilbert's question remains unanswered to this day: the 22 ovals of a curve of degree 8 can be placed on the plane in billions of different ways. But now certain bounds have been found which reduce the number of topologically distinct curves. There are now fewer than 90 cases. However, the number of examples actually constructed, while greater than 70, is not as large as the number of theoretical possibilities.

It is interesting that although the question seems to concern computational mathematics, our computers, so far, have contributed almost nothing to its solution.

If the coefficients of a polynomial are known, then it is possible for a computer to draw the positions of the ovals corresponding to the curve. But a count of *all* possibilities (for any values of the coefficients) is a much more difficult problem.

The problem also has an algorithmic solution, in the sense of mathematical logic. In principle, we can even find the number of connected regions into which the space of polynomials of degree n is divided by a bifurcation diagram, near which the type of the curve changes. But the number of computations needed for this is so large that no progress in computer technology will allow us the hope of a computer solution for the problem of polynomials of degree 8 in the foreseeable future.

Drifting a bit from the theme of today's lecture, I shall talk about one very recent success of computer technology that I know of, with regard to a closely related problem.

Let us think of the graph of a real polynomial of degree n in two variables as a surface, $z = f(x, y)$ in three dimensional space $\mathbb{R}^3$. Near some of its points the surface is locally convex. We call such points *elliptical* points. Around other points, the surface is locally saddle-shaped. We call these points *hyperbolic* points (see Figure 4).

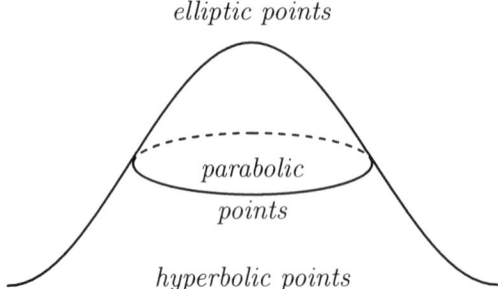

elliptic points

parabolic points

hyperbolic points

Figure 4. The parabolic curve on a smooth surface.

The elliptical and hyperbolic points are divided by a curve consisting of *parabolic points*. In terms of the partial derivatives of the function f, the curve of parabolic points is given by the equation

$$\det \begin{vmatrix} \dfrac{\partial^2 f}{\partial x^2} & \dfrac{\partial^2 f}{\partial x \partial y} \\[2ex] \dfrac{\partial^2 f}{\partial y \partial x} & \dfrac{\partial^2 f}{\partial y^2} \end{vmatrix} = 0,$$

that is, by the condition $f_{xx} f_{yy} = (f_{xy})^2$, or that the Hessian of the function f is zero.

Let f be a polynomial of degree n. We may ask: *how many closed curves (ovals) can its parabolic curve be made up of?*

For a polynomial f of degree 4, the Hessian is also of degree 4, so by Harnack's theorem the number of ovals cannot exceed $g + 1 = 4$.

It is not hard to construct a polynomial of degree 4, with a parabolic curve consisting of three ovals. I leave this problem as an exercise for the reader.

But the problem of whether the parabolic curve of a polynomial of degree 4 can consist of four ovals turns out to be very difficult.

It was solved in Mexico in 2005 by Adriana Ortiz-Rodriguez, who defended her dissertation in Paris as my student. In her dissertation, she proved that the number of ovals in the parabolic curve of a polynomial of degree n is bounded from above by an^2 and below by bn^2, where $a > b$.

When she was still a student (in the University of Paris, Jussieu), she came into my seminar and asked for a problem. I said that to understand my problems, one must first solve the 100 problems of my "Mathematical Trivium" [1]. Good students in Moscow solve them all in 3 hours.

Adriana brought me solutions to these problems, but they all turned out to be wrong. She asked for a week's time to think about them, after which she brought in 10 correctly solved problems. After 10 weeks, she solved all 100 of them and started making sense of mathematics.

But when I wanted to formulate a research problem for her, Adriana said, "No, I thought up a problem of my own, in the style of your seminar and of your students' work on Lagrangian singularities in symplectic geometry." And she formulated the problem which appears above, about parabolic curves.

I replied that I was now convinced that in Mexico they study mathematics as badly as in Paris (where I knew very well how low the level of the students was).

Adriana's inability to solve the problems of the Trivium was the legacy of just this bad training in the basics of mathematics that she had been subjected to, both in Mexico and in Paris. With regard to her imagination and mathematical capability, everything was fine (as her further experience with the Trivium, and with parabolic curves showed). After I taught her everything using my 100 problems, she became an excellent mathematician.

The question of the growth of the number of ovals of a parabolic curve for polynomials of degree n (the question of how the constants a and b in the asymptotic values an^2 and bn^2 from above and below approach each other) remains open to this day. And this is why I included it in this lecture, hoping that I might find some talented students here as well.

As for the original value $n = 4$, Adriana defended her dissertation in Paris, then became a professor in Mexico, where she had unlimited access to computers. In a year of uninterrupted work, the central processing unit of her computer examined 50 million polynomials $f(x, y)$

of degree 4. For three of them, the parabolic curve turned out to consist of four ovals.

When the coefficients of a polynomial are known, the number of ovals in its parabolic curve can be found in a matter of minutes, even without a computer. So the proof of the theorem does not require any mention of a computer experiment. But *finding* these remarkable polynomials without a computer is hopeless, so the application of the computer experiment to the difficult solution of the problem turns out to be decisive.

I hope that my readers will be able to achieve analogous successes in solving the problems we discuss below.

Remark. Before going further, I will explain several things which have been used above, and which are carefully hidden from the students in the traditional pseudo-scientific exposition of mathematics.

The *determinant* of a second-order matrix $\begin{pmatrix} a & b \\ c & d \end{pmatrix}$ is the *area of the parallelogram* constructed using the column vectors (a, c) and (b, d) (see Figure 5), which is written with a plus sign if the vectors orient the plane in the same way as the first and second unit vectors along the axes (and with a minus sign otherwise).

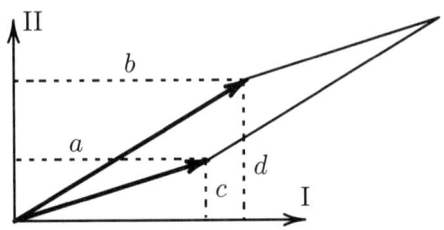

Figure 5. A positively oriented parallelogram.

Two pairs of linearly independent vectors orient the plane *in the same way* if we can connect them with a continuous path in space of ordered pairs of linearly independent vectors on the plane.

There are *exactly two different orientations* (that is, classes of equivalent ordered pairs of vectors on the plane, or ordered frames of n linearly independent vectors, in $\mathbb{R}^n$, for any n). This most important natural scientific fact (which is the only explanation for the strange rule "minus times minus gives plus") is usually kept secret from students, and all this geometry is replaced with the postulate that

$$\begin{vmatrix} a & b \\ c & d \end{vmatrix} = ad - bc,$$

which is, in fact, an easy *consequence* of the topological fact given above. (It is useful also to mention the *linear* dependence of the determinant on each column vector, and its skew-symmetry: the fact that its sign changes when we interchange two columns.)

The second derivatives of a polynomial (or any other smooth function) at a given point form a matrix (of order m for a function of m variables), $\partial^2 f / \partial x_i \partial x_j$. The determinant of this *Hessian* matrix for the function f is simply called the *Hessian* of the function f. It is useful to note that the sign of the Hessian of a function f coincides with the sign of the Gaussian curvature of the graph of the function f (and, by the way, does not depend on the choice of orientation of the space where the function is defined). I will not dwell on this remark because it is meaningful only to those familiar with the sign of the Gaussian curvature, and those to whom it is meaningful can easily prove the assertions made above about the relationship of the Hessian to the Gaussian curvature of a graph.

Let me make one more note, about the genus g of the Riemann surface of an algebraic curve of degree n. We have used, above, the "Riemann-Hurwitz formula":

$$g = \frac{(n-1)(n-2)}{2}.$$

For instance, curves of degree $n = 1$ (a line) and $n = 2$ (a circle) have genus 0; that is, there is a real diffeomorphism between them and the sphere S^2 (also called the Riemann sphere $\mathbb{C} \sqcup \{\infty\}$ or the complex projective line $\mathbb{C}P^1$).

For a line this is clear, and for a circle this follows from the rational parametrization using the "tangent of half an angle" formula

$t = \tan \beta = y/(1+x)$:

$$(1) \qquad x = \frac{1-t^2}{1+t^2}, \; y = \frac{2t}{1+t^2} \quad \text{for } x^2 + y^2 = 1.$$

It is a useful problem to try to understand the topological structure of the "complex sphere" given in projective space $\mathbb{C}P^3$ by the affine equation $x^2 + y^2 + z^2 = 1$. Answer: it is a four-dimensional manifold diffeomorphic to the direct product of two ordinary spheres, $S^2 \times S^2$.

The formulas given in (1) define a diffeomorphism of the complex circle to the sphere S^2. (They also give us the "Egyptian right triangles", $3^2 + 4^2 = 5^2$, $5^2 + 12^2 = 13^2$ and so on, namely, $a^2 + b^2 = c^2$ for $x = a/c, y = b/c$, where, according to the formula (1) with $t = u/v$, $a = v^2 - u^2, b = 2uv, c = u^2 + v^2$.)

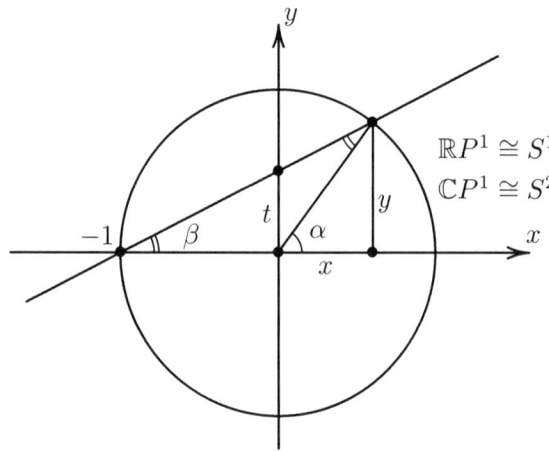

Figure 6. The rational parametrization of a circle.

We can derive formula (1) as follows (Figure 6). We draw the line $\{y = t(x+1)\}$ through the point $(x = -1, \; y = 0)$ in the plane. Substituting this value of y into the equation $x^2 + y^2 = 1$ of the circle, we find the corresponding abscissa x of the point of intersection of the line with the circle by solving a certain quadratic equation, one root of which $(x = -1)$ we already know.

For the other root, Vieta's Theorem (the formula for the sum of the roots of a quadratic equation) gives us a rational expression in t, from which we can derive the first (and then the second) formula of the parametrization (1).

Instead of all this algebra, we can use the geometric identity $\alpha = 2\beta$ from the theorem about the exterior angle (of an isosceles triangle):

$$x = \cos \alpha, \ y = \sin \alpha, \ t = \tan \beta.$$

For those who know some calculus, I note also that from the same rational parametrization of the circle we can get an easy computation (in elementary functions) of any Abelian integral along a circle:

$$I = \int\limits_{x^2+y^2=1} R(x,y)dx,$$

where R is a rational function.

Indeed, the rational parametrization (1) reduces the computation of the integral I to the integration of a rational function of the parameter t,

$$I = \int r(t)dt.$$

Abel proved that *it is impossible to perform this integration in elementary functions if the Abelian integral is taken not along the circle but along some curve of higher genus (with $g > 1$).* For example, this is impossible even for elliptic integrals (along a curve of the form $y^2/2 + U(x) = 0$, where U is a polynomial of degree 3, for example $U(x) = x^3 + ax + b$).

The proof of this topological theorem of Abel is also a marvellous exercise.

It is *topological* because it is not just the function[1]

$$t(X) = \int^X \frac{dx}{y}, \ \text{where} \ \frac{y^2}{2} + U(x) = 0$$

that cannot be presented as a finite combination of elementary functions. In fact such a representation is impossible for any function

---

[1] I.e., a function which expresses the time t it takes to move to a position X under the actions described by Newton's equation: $\dfrac{d^2x}{dt^2} = -\dfrac{dU}{dx}.$

which is topologically equivalent to the (multi-valued) complex function t. And such a representation is also impossible for the inverse functions, which are equivalent to the "elliptic function" $X(t)$ (for non-degenerate value of the coefficients a and b).

The Riemann-Hurwitz formula ($g = (n-1)(n-2)/2$ for smooth curves of degree n) is easiest to prove using the following "Italianate" argument.

Consider any natural family of algebraic-geometric complex objects. An example might be the family of polynomials of degree n in one variable. Other examples are the family of polynomials of a given degree in two variables, which give us algebraic curves, or the corresponding family of homogeneous polynomials of fixed degree in three variables, which give us algebraic curves in the complex projective plane $\mathbb{C}P^2$.

The "Italianate" reasoning consists in noting that *all the non-degenerate objects in this family are topologically identical.* For example, all polynomials of degree n in one variable without multiple roots have the same number of roots; all smooth algebraic curves of degree n in $\mathbb{C}P^2$ have the same genus $g(n)$, independent of the particular choice of a curve.

The proof of this observation is topological. The idea is that the degeneracy of a complex object is determined by a complex equation (the discriminant is equal to zero, in the case of a polynomial in one variable, and so on). And this complex condition on complex coefficients, whose choice gives us an object of the family, gives rise to *two* independent equations in real numbers (both the real and the imaginary parts of the discriminant must vanish).

Therefore *the algebraic variety of all the degenerate objects has a real codimension of two* (in the case of the family of polynomials we're looking at, and so on). But a subvariety with real codimension two cannot divide a smooth variety of all objects of the family into parts (just as a point cannot divide a plane, and a line or curve cannot divide three-dimensional space into parts).

Therefore the variety of non-degenerate objects is connected. From this it follows that all these non-degenerate objects have the same topological type, since if we move along a curve in the space of non-degenerate objects (for example, polynomials without multiple roots), the topological structure of the object (the number of roots of the equation, in the example just given) will not change (by the implicit function theorem).

The principle just demonstrated shows that *to compute the topological characteristics of all non-degenerate objects in a complex family it is enough to look at one example and compute these characteristics for this one object: for all other non-degenerate objects the characteristics will be the same.*

For example, in the case of polynomials of degree n it is enough to take the polynomial

$$f(x) = (x-1)(x-2)\cdots(x-n),$$

which obviously has the n roots $x = 1, 2, \ldots, n$ (each with multiplicity 1).

By the "Italianate Principle", a consequence of this observation is the "Fundamental Theorem of Algebra": *every polynomial of degree n in one variable without multiple roots has exactly n complex roots.*

In the case of planar algebraic curves it is sufficient to find the genus of one (non-singular) curve of degree n.

Let us begin the topological investigation with a singular curve of degree n, which breaks into n lines, intersecting in pairs at $n(n-1)/2$ distinct points (Figure 7).

If the equation of this curve has the form $f_0 = 0$, where f_0 is the product of n linear non homogeneous functions $a_j x + b_j y + c_j$ then the equation $f_\varepsilon = 0$ (where $f_\varepsilon = f_0 - \varepsilon$) gives us a smooth curve of degree n for small values of $\varepsilon \neq 0$. We now compute the genus of this curve.

This computation proceeds as follows. The curve $f_0 = 0$ is formed by n spheres S_j^2, which intersect in pairs at $n(n-1)/2$ distinct points. For small ε, the transition to the curve $f_\varepsilon = 0$ requires changing the cross formed by two transversally intersecting smooth spheres near their point of intersection into a cylinder connecting the completions of neighborhoods of the points of intersection on each sphere (Figure 8).

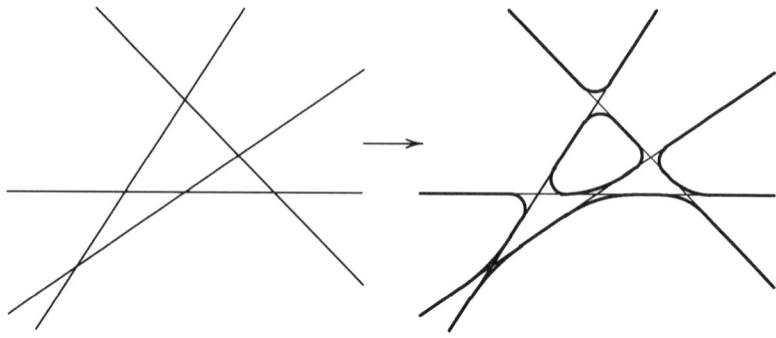

Figure 7. A deformation of a decomposable curve.

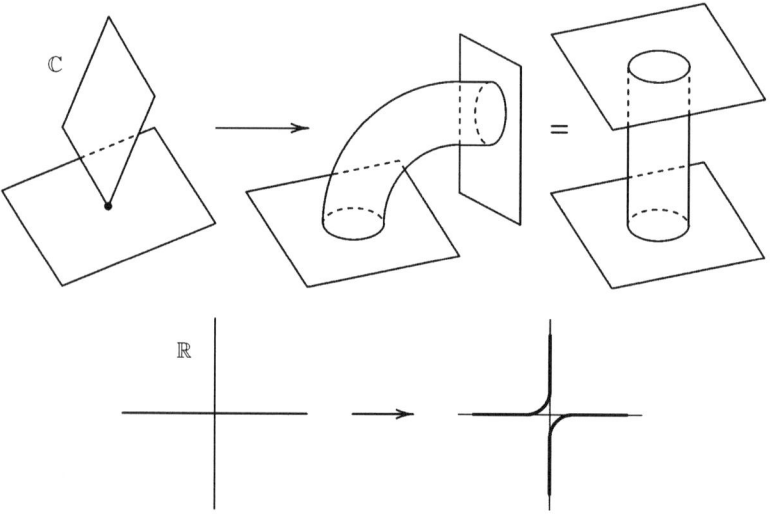

Figure 8. The deformation at a double point.

Let us find out how many handles are created after $n(n-1)/2$ such independently completed deformations (near each point of intersection).

Let us take one of the n spheres $S_1^2, \ldots, S_n^2$, say S_1^2. After our deformation, its $n-1$ points of intersection with the other spheres connect each of the remaining spheres with the first, so that together they form another surface Σ^2 which is again diffeomorphic to the sphere, except that we haven't taken into account the remaining $(n-2)+(n-3)+\cdots+1 = (n-1)(n-2)/2$ points of intersection of the other spheres with each other (Figure 9).

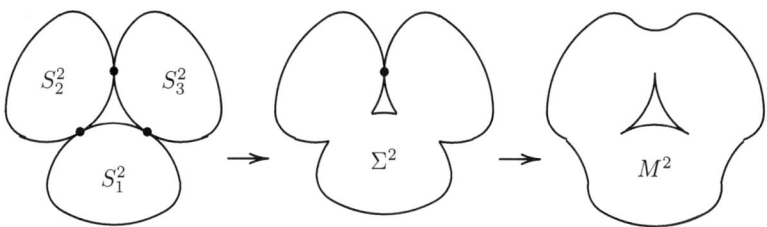

Figure 9. Constructing a surface from a set of spheres.

After transforming the $(n-1)(n-2)/2$ points of self-intersection of the surface Σ^2 using the same number of tubes, we turn the "sphere" Σ^2 into a (smooth) surface M^2 which is a sphere with handles. The number of handles g is the number of tubes which replaced the points of self-intersection of the "sphere" Σ^2, so $g = (n-1)(n-2)/2$.

Figure 9 shows the case $n = 3$, where we get $g = 1$. Hence the surface M^2 turns out to be a torus, which is of genus 1.

Thus we obtain the Riemann-Hurwitz formula $g = \dfrac{(n-1)(n-2)}{2}$.

Harnack's inequality, the assertion that a real curve of genus g has no more than $g+1$ ovals, turns out to be a special case of *Smith's inequality*:

$$(2) \qquad \sum b_k(M_{\mathbb{R}}) \leq \sum b_k(M_{\mathbb{C}}).$$

Here, $M_{\mathbb{C}}$ is a complex algebraic variety (for example, the Riemann surface of a curve). If this variety is given by an equation with real coefficients, then it is acted on by the symmetry ("involution") $\sigma : M_{\mathbb{C}} \longrightarrow M_{\mathbb{C}}$ of complex conjugation (the replacement of a point with complex coordinates $z_j = x_j + iy_j$ by the point with complex coordinates $\overline{z_j} = x_j - iy_j$). Obviously, $\sigma^2 = 1$, and the real variety $M_{\mathbb{R}}$ consists of the fixed points of the involution σ (which are ovals, in the case of a curve).

The numbers b_k in inequality (2) are "the Betti numbers" for the chains with coefficients in the group $\mathbb{Z}_2$ of two elements.

The Betti numbers for the circle are

$$b_0 = b_1 = 1, \quad b_k = 0 \text{ for } k > 1.$$

For a Riemann surface of genus g we have

$$b_0 = b_2 = 1, \quad b_1 = 2g, \quad b_k = 0 \text{ for } k > 2.$$

Here the $2g$ one-dimensional cycles are the "parallels" and "meridians" of the g handles.

Thus, in the case of curves of genus g, Smith's inequality takes the form

$$2 \text{ (number of ovals)} \leq 2g + 2;$$

that is, we have obtained Harnack's inequality:

$$\text{number of ovals} \leq g + 1.$$

The proof of Smith's inequality itself is not that difficult if we look at the action of the involution σ on all possible chains (of a triangulation of the manifold which is symmetric with respect to the involution σ). In the case of Riemann surfaces of real curves the most important argument of Smith's theory lies in the fact that among its ovals there can exist *only one* homological relationship (the sum of the ovals is homological to zero) in the one-dimensional homology of the Riemann surface: otherwise this surface would not be connected, but would fall into connected two-dimensional components (made up of pieces of the form a and σa, where the boundary of the two-dimensional chain a is the left part of the relation between the ovals).

To these remarks from real algebraic geometry I can add that in addition to graphs of polynomials, Adriana Ortiz-Rodriguez, in her dissertation, also examined parabolic curves on any algebraic surface of degree n in the real three-dimensional projective space $\mathbb{R}P^3$. In this case, the number of parabolic curves is bounded by her from above and below by the quantities an^3 and bn^3, where the constant a is approximately 10 times greater than the constant b.

I formulate this result because I have hopes that my audience might want to find the exact rate of growth of the number of parabolic curves, by moving a and b closer together.

The results of Gudkov about curves of degree 6 have laid the foundation for a remarkable new theory connecting the real algebraic geometry of Hilbert's 16th problem with quantum field theory and with multi-dimensional topology.

The fact that the number of interior ovals in Gudkov's list of curves of degree 6 with 11 ovals (1, 5 and 9) increases by 4 is not accidental. Passing from the curve $f(x, y) = 0$ to the bounded surface with boundary M: $f(x, y) \geq 0$, we come to a sequence of Euler characteristics which differ by 8.

Analyzing the real projective algebraic curves of degree $n = 2k$ found by Gudkov which have the largest possible number of ovals according to Harnack's theorem, I noticed that for these curves, the Euler characteristics of the surfaces M satisfy the congruence

$$(3) \qquad\qquad \chi(M) \equiv k^2 \pmod 8,$$

which I call "Gudkov's congruence". (See Editor's note 2, page 56.)

Congruences modulo 8 (for the signatures of intersection forms) turn out to be standard in the topology of four-dimensional closed varieties, so I started searching for four-dimensional manifolds in the topology of (*one-dimensional*) real algebraic curves.

These manifolds turned out to be *complexifications of surfaces with boundaries* M^2. To complexify a surface given by an inequality $f(x, y) \geq 0$, I write down this geometric inequality in the algebraic form $f(x, y) = z^2$. In the complex domain, this formula defines a manifold of real dimension four: a two-sheet covering of the complement of the Riemann surface (of the complex curve) $f(x, y) = 0$ in $\mathbb{C}P^2$, branching along this Riemann surface.

Applying topological results about the divisibility by 8 of the signature to this four-dimensional variety, I proved congruence (3) modulo 4. Then Rokhlin, using deeper results from differential topology of smooth four-dimensional varieties, proved Gudkov's congruence (3) itself.

It is interesting that Gudkov himself, whom I informed of this congruence while writing a report on his dissertation, thought that it was false, since he believed he knew counter-examples to it (which, however, turned out to be as false as Hilbert's results about curves of degree 6, which had been disproved in this very dissertation.)

Recently, congruence (3) has become the foundation for a large number of new results in real algebraic geometry, in differential topology, and even in quantum field theory. But, unfortunately these results are not sufficient even for the classification of topological structures of curves of degree 8 in Hilbert's 16th problem.

Returning to the 16th problem, I note that Hilbert, in my view, left out the most important questions in formulating it.

The point is that topological structures can be different not only for real algebraic curves (of a given degree) $\{f(x, y) = 0\}$, but also for polynomials $f\colon \mathbb{R}^2 \to \mathbb{R}$ defining these curves.

In formulating his problem, Hilbert should have included not just the question of topological classification of curves of a given degree in the real projective plane, but also the question of the *topological classification of the polynomials defining these curves.*

This problem has not been solved, as far as I know, even for curves of degree $n = 4$ (which Descartes had already classified). In the next section I will discuss the topological classification of functions and polynomials, where much remains unknown (which is partly Hilbert's fault).

2. The Statistics of Smooth Functions

To describe the topological structure of a smooth real function, we associate it with the graph whose points are the connected components of the level hypersurfaces of this function.

For a non-degenerate "Morse Function" $f\colon S^n \to \mathbb{R}$, $n > 1$ (see Editor's note 3, page 56), this graph turns out to be a tree with T triple branch points, $K = T+2$ end vertices or 'leaves' and $P = 2T+1$ edges that connect $K + T = 2T + 2$ vertices of the graph.

Example 1. For the "Mount Elbrus Function" (which shows the altitude at each point on the mountain), there are two local maxima,

A and B, and one saddle point C. Hence the graph looks like the letter Y (Figure 10).

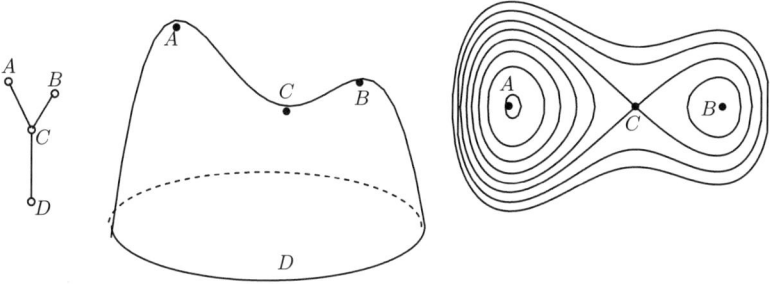

Figure 10. Graph and level curves for Mount Elbrus.

We will consider functions $f\colon \mathbb{R}^n \to \mathbb{R}$ that behave like $-r$ far from the origin, and extend them near the point $\infty = S^n \setminus \mathbb{R}^n$ so that this point is a local minimum (the end vertex D of the tree above).[2]

Example 2. For Mt. Vesuvius we have Figure 11.

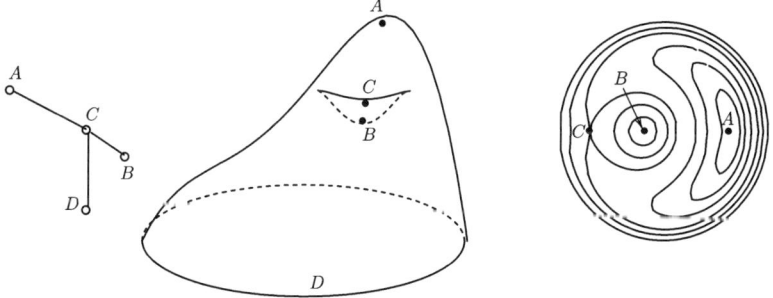

Figure 11. Graph and level curves for Mount Vesuvius.

In constructing our graph for each function we will include the ordering of the vertices by height $(f(A) > f(B) > f(C) > f(D)$ for

---

[2]where r is the distance from the origin. (transl.)

Elbrus; $f(A) > f(C) > f(B) > f(D)$ for Vesuvius), which distinguishes the graphs of example 1 and example 2, even though these trees are homeomorphic. (See Editor's note 4, page 56.)

For the sake of simplicity, we assume that all the $2T + 2$ critical values of the function f are distinct. For functions on the sphere S^2, T of these values correspond to saddle points, while $T + 2$ correspond to maxima and minima. The graphs of the Morse function on the sphere S^n, $n > 2$ resemble the graphs for $n = 2$ and are also trees, but we will later apply techniques we develop in studying these trees to the case of functions on the torus $T^2 = S^1 \times S^1$, where the graphs may contain cycles.

Our basic goal will be to study the statistics of the graphs of functions (whose vertices are ordered by the heights of the summits), having T triple nodes: which of these ordered trees are the graphs of polynomials with degree corresponding to the value of T?

A typical polynomial of degree n in two variables has no more than $(n-1)^2$ critical points on the plane $\mathbb{R}^2$. This corresponds to the value $2T + 2 = (n-1)^2 + 1$; that is, $T = 2k(k-1)$ triple nodes of the graph (the saddle points of the function) for polynomials of degree $n = 2k$. (See Editor's note 5, page 57.)

Theorem 1. *For $T \leq 4$, the number $\varphi(T)$ of (ordered) graphs of functions (trees with T triple vertices) is*

T	1	2	3	4
$\phi(T)$	2	19	428	17746

The ordering of the graphs of these functions has the following property: among the three neighbors of any branch point there are both vertices that are higher than this point (1 or 2 of them) and lower than this point (2 or 1 of them). Ordering with this property will be called *proper*.

This follows from the fact that the graphs of Elbrus and Vesuvius (Figures 10 and 11) are properly ordered, and that the topological structure of a function in the neighborhood of a saddle point is always the same as that of either Elbrus or Vesuvius.

The following two results give estimates of the rate of growth of the number $\varphi(T)$ of properly ordered graphs as T increases.

Theorem 2. *The number $\varphi(T)$ of properly ordered trees (graphs of functions) with T triple vertices has the following lower bound*

$$\varphi(T) \geq \frac{(T^2 + 5T + 5)(2T + 2)!}{(T + 4)!}.$$

For $2 \leq T \leq 4$ the right-hand side of this inequality has, respectively, the values 19, 232, 3690. The lower bound for the growth of the right hand side obtained from Stirling's formula gives a value of $4(4/e)^T T^T > T^T$.

Theorem 3. *The number $\varphi(T)$ of properly ordered trees with T triple vertices has the following upper bound:*

$$\varphi(T) \leq T^{2T}, \quad \text{for } T > 2.$$

The proof of Theorem 2 is based on a direct count of the number of those special ordered trees for which the T triple points form a monotone A-chain with critical values

$$f(A_1) > f(A_2) > \cdots > f(A_T)$$

at adjacent vertices of the graph $A_1 - A_2 - \cdots - A_T$.

From now on, we will assume that the functions whose graphs we examine are defined on this graph as well: the value of a new function at each point of the graph is equal to the value of the original function at each point of the level hypersurface, whose connected component corresponds to this point of the graph (again see Editor's note 3, page 56).

Theorem 4. *The number of properly ordered graphs that are trees with T triple points forming an ordered A-chain is equal to*

$$\psi(T) = \frac{(T^2 + 5T + 5)(2T + 2)!}{(T + 4)!}.$$

Theorem 2 follows from Theorem 4, because the number $\varphi(T)$ of all proper orderings is no less than the number $\psi(T)$ of proper orderings in which the triple points form a monotonic A-chain.

Remark. Some of our properly ordered graphs are graphs of *polynomials* (of degree $n = 2k$ for $T = 2k(k-1)$), and some are not.

It would be interesting to know if the number of proper orderings given by polynomials is small compared to the number of orderings given by all smooth functions. Or perhaps the opposite is true: the number of proper orderings *not* given by polynomials is relatively small (asymptotically, as $T \to \infty$).

If a graph is given by a function, the number of topologically distinct ways in which this can be done is also of interest. Here we must also study the question of the topological classification of topologically different smooth Morse functions, and the question of the number of connected components of the space of polynomials with properly ordered graphs of a given degree (which may turn out to be greater than one, even in the case where all these polynomials are topologically equivalent).

Both questions remain open, and I expect accomplishments from the audience.

Proof of Theorem 4. The transition from the chain $A_1 - A_2 - \cdots - A_T$ to the full graph consists in adding two edges to each of the vertices A_1 and A_T and one edge to each of the vertices $A_2, \ldots, A_{T-1}$. We will call this process a *charging*. Such chargings differ from each other by the ordering of heights of the free ends of the additional edges. We need to count the number of different chargings.

Let us denote by a and α the end vertices adjacent to the triple vertex A_1 and different from A_2. We assume that $f(a) > f(\alpha)$. Thus $f(a) > f(A_1)$.

Similarly, we denote by z and ω the end vertices adjacent to the triple vertex A_T and different from A_{T-1}. Again, we assume that $f(z) < f(\omega)$ and $f(z) < f(A_T)$. (See Figure 12.)

To classify the chargings of the triple vertices $A_1, \ldots, A_T$, we note first that the critical value $f(\alpha)$ belongs to the complement of

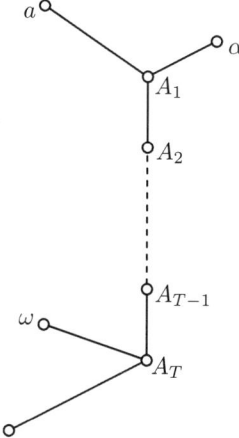

Figure 12. Charging the end vertices A_1 and A_T of a chain.

the following set of $T + 1$ real numbers smaller than $f(\alpha)$:

$$\{f(A_1), \ldots, f(A_T); f(z)\}.$$

It follows that there are $T+1$ topologically distinct cases in which $f(\alpha) > f(z)$, and one more case, topologically different from all of the others, when $f(\alpha) < f(z)$.

Knowing the interval in which the critical value $f(\alpha)$ falls, we consider the critical value $f(\omega) > f(z)$. It must be different from the $T + 2$ values

$$\{f(A_1), \ldots, f(A_T); f(a), f(\alpha)\},$$

if $f(\alpha) < f(z)$, and must be different from the $T + 1$ values

$$\{f(A_1), \ldots, f(A_T); f(a)\},$$

if $f(\alpha) < f(z)$.

Hence we find

$$(T + 1)(T + 3) + 1(T + 2) = T^2 + 5T + 5$$

topologically distinct chargings (α, ω) *of triple vertices* A_1 *and* A_T.

For the endpoint a_2 of the graph which is connected to the triple vertex A_2, the critical value must be different from the $T + 4$ values already chosen,

$$\{f(A_1), \ldots, f(A_T); \; f(a), \; f(z), \; f(\alpha), \; f(\omega)\},$$

which subdivides each of the previous cases into $T + 5$ subcases. After choosing $f(a_2)$, we obtain $T + 5$ prohibited values for $f(a_3)$, and so on. For $f(a_i)$ we have to avoid $T + 2 + i$ previously chosen values,

$$\{f(A_1), \ldots, f(A_T); \; f(a), \; f(z), \; f(\alpha), \; f(\omega); \; f(a_2), \ldots, f(a_{i-1})\}.$$

These values subdivide the real axis into $T + 3 + i$ intervals.

Repeating this reasoning $T - 2$ times (for $i = 2, \; 3, \; \ldots, \; T - 1$), we subdivide each of the $T^2 + 5T + 5$ cases for the chargings of the end triple vertices A_1 and A_T into many subcases. The number of these subcases is equal to the product of the numbers of intervals in the consecutive steps of our construction, or

$$(T + 5)(T + 6) \cdots (T + 3 + T - 1) = \frac{(2T + 2)!}{(T + 4)!}.$$

These subclasses give us all the topologically distinct chargings, and each charging appears exactly once. This proves Theorem 4.   □

To prove Theorem 3, we start with the following (inductive) proposition.

Lemma. *For any $T \geq 2$ we have the following inequality:*

$$\varphi(T) \leq 4T^2 \varphi(T - 1).$$

Example. For $T = 2, 3$, and 4 we have, by direct computation:

$$(\varphi(2) = 19) < (16 \cdot 2 = 32);$$
$$(\varphi(3) = 428) < (36 \cdot 19 = 684);$$
$$(\varphi(4) = 17746) < (64 \cdot 428 = 27392);$$

showing that the assertion of the Lemma is true for these cases.

Proof of Lemma. The maximal critical value is achieved at one of the end vertices A of a connected graph T with triple vertices. This end vertex is connected by an edge to some triple vertex B. Erasing the edge AB, we reduce the original graph with T triple vertices to a smaller related graph with $T - 1$ triple vertices.

There are $\varphi(T-1)$ of such smaller (properly ordered) graphs. To recover the original larger (properly ordered) graph, we must take some edge in the smaller graph, place a new triple vertex B on it, and draw an edge connecting it to a new end vertex A that is higher than all the others.

The smaller graph has $2T-1$ edges (see Editor's note 6, page 57), and the vertex B must belong to one of them The value at the chosen vertex B must be distinct from the $2T$ other values at the vertices of the smaller graph. This gives $2T+1$ different options (different topological types of graphs).

The total number of both choices is equal to $(2T-1)(2T+1) = 4T^2 - 1 < 4T^2$, which implies the inequality in the Lemma. □

In fact we have proven more than the assertion of the Lemma. We have found an upper bound not just for $\varphi(T)$ (the number of graphs of functions). We have found a larger number, which also counts the "improperly" ordered trees, in which the altitude of some triple vertex is greater than all three altitudes of its neighboring vertices (or smaller than all three). This cannot happen in a properly ordered graph of a function, the number of which is therefore smaller than the number we have found.

Proof of Theorem 3. For $T = 3$ we have

$$(\varphi(3) = 428) < (3^6 = 729).$$

If the inequality of Theorem 3 is true for $T = S - 1$, the lemma gives us the inequality

$$\varphi(S) \leq 4S^2(S-1)^{2S-2}. \qquad (*)$$

Applying the obvious inequality

$$\left(1 - \frac{1}{S}\right)^S < \frac{1}{e},$$

we get an estimate for the right hand side of inequality $(*)$:

$$4S^2(S-1)^{2S-2} \leq \frac{4S^2}{(S-1)^2}\left(\frac{S-1}{S}\right)^{2S} S^{2S} \leq \frac{4}{e^2}\frac{S^2}{(S-1)^2}S^{2S}.$$

The coefficient $4S^2/e^2(S-1)^2$ is less than 1 when $2S \leq e(S-1)$, which is true when $S \geq 4$.

In other words, if $S \geq 4$, then the inequality of Theorem 4 for $T = S - 1$, together with inequality $(*)$ gives us the inequality $\varphi(S) \leq S^{2S}$ of Theorem 3, which therefore is proved for $T = 4, 5, \ldots$.   □

Proof of Theorem 1. Consider T triple vertices of a tree with $2T + 2$ vertices. They are the set of vertices of a connected subgraph, which is obtained from the original graph with T triple points by deleting $T + 2$ edges connecting them with end vertices.

For $T = 1, 2, 3$ the remaining (ordered) graphs are shown in Figure 13.

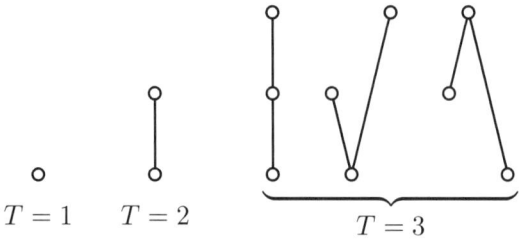

Figure 13. Shortened forms of ordered trees with $T \leq 3$ triple vertices.

In the first three cases, the chargings can be counted as in the proof of 3. This gives us:

$$\varphi(1) = 2, \quad \varphi(2) = 4 + 5 \cdot 3 + 5 = 19,$$
$$\psi(3) = (9 + 5 \cdot 3 + 5) \cdot 8 = 232$$

chargings, respectively.

The analysis of each of the remaining two cases can proceed by analogy with the proof of Theorem 4 above, giving 98 chargings in each case.

The equality of the number of chargings of these two ordered graphs is obvious *a priori* from the symmetry (reflection in the horizontal axis) taking each of these ordered graphs into the other.

For $T = 4$ we must consider 9 essentially different cases (Figure 14).

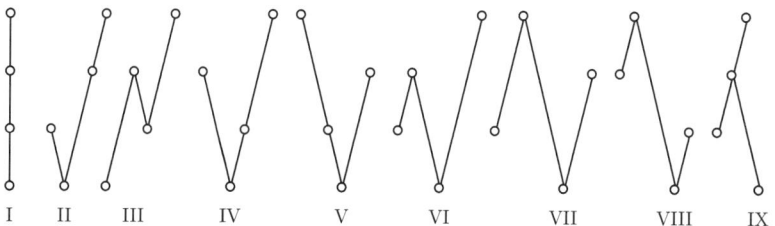

Figure 14. Shortened forms of ordered trees with $T = 4$ triple vertices.

Each of cases II, IV, VI, IX gives us two ordered graphs, which are symmetric with respect to the horizontal axis (like the last two graphs analyzed earlier, for $T = 3$).

These symmetric graphs are different, but they have the same number of chargings. (We can see this from the symmetry, extended to the symmetry of the chargings of symmetric graphs).

The number ψ of chargings for these 9 cases is given by the following table:

case	I	II	III	IV	V	VI	VII	VIII	IX
ψ	3690	1680	586	1360	1360	534	486	756	1180

Case I has the largest number, and was analyzed in the proof of Theorem 2. The count of the chargings for the other cases follows the same method, but the details are too numerous to give in this lecture. I would rather think of it as a natural exercise for a formal course.

The total number of chargings for all the shortened graphs, including symmetric pairs of cases, is given by the following sum:

$$\varphi(4) = \psi(\mathrm{I}) + \psi(\mathrm{III}) + \psi(\mathrm{VII}) + \psi(\mathrm{VIII})$$
$$+ 2\psi(\mathrm{II}) + \psi(\mathrm{IV}) + \psi(\mathrm{V}) + \psi(\mathrm{VI}) + \psi(\mathrm{IX})$$
$$= 5518 + 2 \cdot 6114 = 17746,$$

which also proves Theorem 1 (for $T = 4$). $\qquad\qquad\square$

I don't know if the case similar to case I gives the largest number of chargings for each $T > 4$.

Remark. It would be interesting to analyze how $\varphi(T)$ grows as T increases. It is likely that it grows essentially as T^{cT} (up to a "log-arithmic" correction like const^T). The constant c (which is between 1 and 2, according to Theorems 2 and 3) seems empirically closer to 2 (and may even be equal to 2, up to a multiplicative "logarithmic" constant).[3]

The computations above of the functions ψ for nine cases ($T=4$) indicate a certain homogeneity in the distribution of all charged graphs among the nine classes described. For example, for the special type of subgraph with T triple vertices that have critical values

$$\mathrm{II}(T) = \{a_1 > a_2 > \cdots > a_{T-1}; \ a_1 > b > a_2\}$$

at the sequence of vertices $(B, \ A_1, \ A_2, \ \ldots, \ A_{T-1})$ the asymptotics of the numbers of assignments of weights is such that

$$\frac{\psi(\mathrm{II}(T))}{\psi(\mathrm{I}(T))}) \to \frac{1}{2} \text{ as } T \to \infty, \text{ so that } \psi(\mathrm{II}(T)) \sim \frac{T^2 \cdot (2T+2)!}{2 \cdot (T+4)!}.$$

It would be interesting to calculate the corresponding ratio between the asymptotics of the numbers of chargings for other reduced graphs. How do they depend on the geometry of the reduced ordered graphs and why does the ratio $1/2$ occur in the case of the pair I, II? Does the corresponding ratio always approach a rational number? Are these numbers less than one?

I even hope that such an investigation will result in the lower bound

$$\varphi(T) \geq BT^2 \varphi(T-1),$$

for some constant B, and even, perhaps, for $B = 2$.

The following (non-rigorous) "physical" argument prompts this hope. We distinguished above $4T^2 - 1$ ways to add a new edge leading to the highest critical point. The majority of these ways in fact give us new graphs with one new triple vertex, chosen arbitrarily from $2T - 1$ edges of the smaller graph with $T - 1$ triple vertices. The only difficulty lies in the choice of a critical value for the new vertex. For this choice we lay out $2T + 1$ intervals, but the new critical value w

---

[3]L. Nicolaescu proved this conjecture in 2006 (see [**16**]), after learning about the present lecture in 2005.

should not be smaller than the values u and v at the end vertices of the edge at which we have chosen the new vertex.

The trouble is that (see Figure 15) the third neighbor of the new triple vertex is the highest vertex of the new ordered graph. Therefore, if $u > w$ and $v > w$, then the critical value w turns out to be smaller then each of the three critical values in neighboring vertices of the graph (which never happens in the ordered graph of a good Morse function).

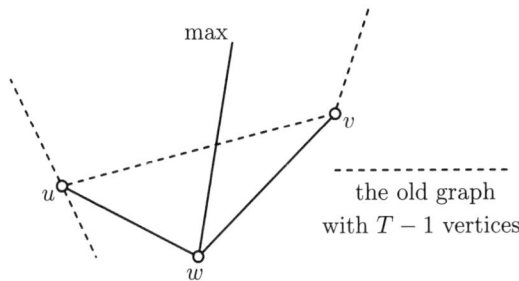

Figure 15. An impossible choice for a new critical value w.

Let us try to estimate how much this obstacle decreases the number of larger graphs that can be constructed from a given shortened graph.

A heuristic (non-rigorous) argument is hinted at by the "probability" of the event $\{w < u,\ w < v\}$, which is of order $1/4$, since each of the inequalities apparently has "probability" $1/2$, and they seem to be independent.

If this is so, then we must omit one quarter of the full number $4T^2 - 1$ of larger graphs for which there is no corresponding function. The asymptotic "ergodicity" used in this non-rigorous proof (the distribution of values at the vertices of a randomly chosen graph) may be not easy to prove. Aside from proving it, one can investigate it empirically with an experimental count of all the graphs (say, with $T = 5, 6, 7, 8$ triple vertices), for which we would need just a few hours of computer time.

Having accepted the "probability" $1/4$, we can change the coefficient $B = 4$ to $B = 3$. But the numbers in Theorem 1 hint at an even smaller value of the bound (as T approaches infinity) of the ratio

$$\frac{\varphi(T)}{T^2 \varphi(T-1)},$$

which is closer to 2 than to 3.

This smaller value of the coefficient B can be explained (at least heuristically) by the following obstacle to the construction of a larger ordered graph. (See Figure 16.)

Suppose that the new value w is higher than the values at either end of the edge on which a new triple vertex is chosen: $w > u, w > v$.

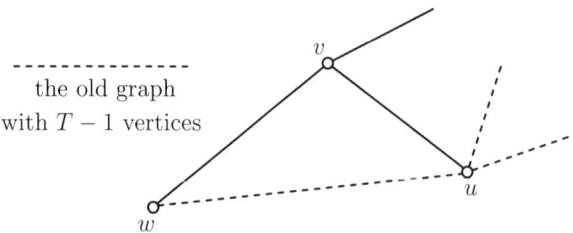

Figure 16. An impossible choice for a new critical value w.

If the three values are given in the order

$$v < u < w,$$

then the choice of the value w on the new triple vertex might ruin the old triple vertex, where we have the value u. This will happen if the values at both vertices neighboring this one in the old graph are greater than u. In this situation, the value u in the new ordered graph (on the old triple vertex of the new graph) will be less than the values on each of the three neighboring vertices (in the new graph). This is impossible for a Morse function.

In re-computing the "probability" of a new obstacle, we must replace the coefficient $B = 3$ with three-quarters of its value, to get $B = 9/4$. This (semi-empirical) proposition is not far from the values we have observed. The ratio

$$\frac{\varphi(4)}{\varphi(3)} = \frac{17746}{428}$$

is not very different in value from the product

$$\frac{B}{T^2} = \frac{9/4}{4^2} = 36.$$

Of course, these semi-empirical conclusions must be supported not only by a proof (which may be not at all simple) of the hypothetical assertion formulated above of the ergodic theory of random graphs (leading to $B = 9/4$), but also by numerical experimentation (for which we must either compute several consecutive values of $\varphi(T)$, or else count directly, or perhaps just display random orderings of random trees, which may require less computer time).

No matter how we proceed, and no matter what constant B gives us the inequality

$$\varphi(T) > BT^2\varphi(T-1),$$

it should lead us to a rate of growth of $\varphi(T)$, similar to the growth of the quantity T^{2T} (neglecting a factor of the form const^T which depends on the coefficient B and is "logarithmically small" in comparison to T^{2T}).

In fact an upper bound in Theorem 3 is proven above for a number of ordered graphs bigger than $\varphi(T)$, including graphs unrealizable as ordered graphs of functions. (Unrealizable graphs allow triple vertices that are higher than all three neighboring vertices in the graph, or lower than all three. This cannot happen in the ordered graph of a function.)

To obtain a lower bound from the proof for the upper bounds given above, we would have to know the "probabilities" of the unpleasant situations described. However, calculating these probabilities requires difficult results from the ergodic theory of the random orderings of random graphs which I cannot obtain (either by proving them, or by verifying them with numerical experimentation).

It is exactly in the hopes that my listeners will succeed in this new and unexplored area that I have included above the semi-empirical theory in this lecture.

Aside from the topological classification of ordered graphs of functions, the topological classification of the functions themselves (with a given graph Γ) on a fixed manifold M is also interesting.

In the case of a two-dimensional manifold (for example, the sphere S^2), an ordered graph (with given critical values) apparently[4] defines a topological type of a Morse function. In the case of higher dimensions the situation is more complicated, and the appropriate version of our combinatorial theory has not been developed. Even if we fix the Morse indices and the critical values corresponding to the vertices of the graph, it is not clear how many types of Morse functions on S^3 correspond to a given graph.

The function f on M is obtained from a function $\widetilde{f}$ on its graph Γ by using the natural mapping $\pi : M \to \Gamma$, which assigns to each point of the domain of definition of f the component of the set of levels containing this point.

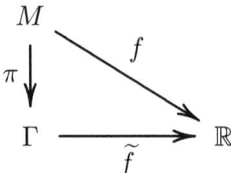

Topological invariants (for example, the homotopy invariants) of the natural mapping π provide interesting topological invariants of a function f (with a given graph Γ). It would be interesting to know what they are (and what mappings π are realized by smooth functions f on a given manifold M).

For example, this question, which seems easy for $M = S^2$, is more interesting for trigonometric polynomials in two variables, $f : T^2 \to \mathbb{R}$. The corresponding graph Γ has one cycle (g cycles, for a surface M^2 of genus g). The proof of this property of the graph Γ is a useful exercise in the calculation of the Euler characteristic:

$$\chi(M^2) = 2 - 2g,$$

$$\chi(\Gamma) = \text{(number of vertices)} - \text{(number of edges)}$$
$$= (T + K) - (3T + K)/2$$
$$= 1 - \text{(number of cycles in } \Gamma).$$

---

[4] A rigorous proof seems not to have been published yet. It should be.

For a trigonometric polynomial f with a given degree n the topological complexity of the transformation $\pi : T^2 \to \Gamma$ must clearly be bounded by some function of n (not yet found). It would be interesting to know how many different homotopy classes of transformations π are realized by trigonometric polynomials of a given degree n (or with a given spectrum, which is a finite subset of the lattice of free vectors $\mathbb{Z}^2$).[5]

3. Statistics and the Topology of Periodic Functions and Trigonometric Polynomials

The Morse function $f : T^2 \to \mathbb{R}$ with T critical saddle points and K maximum and minimum points has a graph with T (triple) branch points and $K = T$ endpoints. This graph (see Figure 17) has $P = 2T$ edges and one cycle (with trees adjoined to its points).

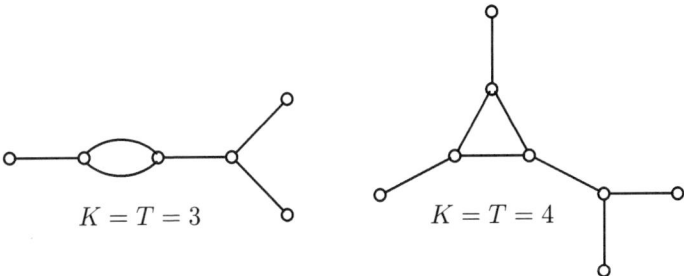

$$K = T = 3 \qquad\qquad K = T = 4$$

Figure 17. Some graphs with one cycle and T triple branch points.

Example. Consider a four-parameter family of trigonometric polynomials

$$f_{A,B,C,D}(x, y) = A \sin x + B \sin y + C \sin(x + y) + D \cos(x + y). \quad (1)$$

We can think of these as functions on a torus:

$$T^2 = \{x \,(\mathrm{mod}\,2\pi), y \,(\mathrm{mod}\,2\pi)\}$$

---

[5]The first results along these lines were published in 2006; see **[6]**, **[11]**.

As we will see, the largest number of critical points they can have (for a non-degenerate Morse function) is

$$K = T = 4.$$

Thus the numbers of vertices and edges are

$$V = 8, \quad E = 8.$$

(In fact the maxima are smaller: $K = T = 3$.)

A count shows that *the number of directed graphs of smooth functions on a torus with these values of the parameters is 550, and the polynomials in the family (1) provide no more than 12 of these 550 properly ordered graphs.*

Hypothetically, if we increase the degree of the trigonometric polynomials, their graphs should constitute a smaller and smaller part of the set of all properly ordered graphs with that many vertices, and I hope that my audience will help move us towards a proof of this hypothesis.

Aside from the classification of ordered graphs of trigonometric polynomials of a given degree, the topological classification of the polynomials themselves is also of interest. Moreover, even topologically equivalent trigonometric polynomials of a given degree or with a given spectrum can form several connected regions of the space of all such trigonometric polynomials. The study of the topology of these regions (in particular, their number) is an interesting question, which I mention here in the hopes that my listeners will play a role in its solution.

Aside from the set of all trigonometric polynomials of a given degree, it is interesting to study the more general space of trigonometric polynomials with a given spectrum S,

$$f(z) = \sum_{k \in S} f_k e^{i \langle k, z \rangle}, \quad f_k \in \mathbb{C}, \quad z \in \mathbb{C}^m.$$

Here, the "wave vectors" $k \in S \subset \mathbb{Z}^m$ of harmonic waves on an m-dimensional torus belong to the finite "spectrum" S. For the formula to yield a real trigonometric polynomial $f : T^m \to \mathbb{R}$, the coefficients must satisfy the condition that they be real ($f_{-k} = \overline{f}_k$).

Recall that the notation $\langle \cdot, \cdot \rangle$ used above denotes the Hermitian inner product in the space $\mathbb{C}^m$ (given by the formula $\langle k, z \rangle = \sum_{j=1}^{m}(k_j \overline{z}_j)$ for vectors k and z with components k_i and z_i, where $\overline{z} = x - iy$ for $z = x + iy$; that is, the "overbar" represents complex conjugation).

I assume that the audience is familiar with the *basic theory of harmonic waves* from a school course in physics (or from the elementary text of Landau and Lifshitz), so that they will understand that the following complex formulas can be written in the real form

$$\sum_k (a_k \cos(k, x) + b_k \sin(k, x)),$$

where $x \in \mathbb{R}^m$, $k \in S$, and $(k, x) = \sum(k_j x_j)$ is the Euclidean inner product.

We can get a set of trigonometric polynomials with particularly interesting spectra by taking the "extended root systems of affine groups of reflections" (which I will not describe here). I note only that for the simplest group A_2 (corresponding to the symmetries of an equilateral triangle), the corresponding spectrum consists of the six vectors $\{\pm\alpha, \ \pm\beta, \ \pm(\alpha + \beta)\}$ in the two-dimensional plane.

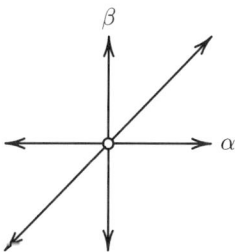

The trigonometric polynomials corresponding to this spectrum are of the form (1). They form a four-parameter family, rather than a six-parameter family: we would need combinations of the functions $\cos x$ and $\cos y$, but can do without these by choosing a certain point of the torus T^2 as the origin of the coordinate system. In this way, the general case of a function from the six-parameter family can be reduced to case (1), simply by changing the origin.

Transferring the theory outlined below for the case of the simplest family (1) to the case of trigonometric polynomials of higher degree (or those related to more general systems of roots, or even with any spectra) is one of the reasons for including the elementary theory described below in the present lecture.

Here we can doubtless make quick progress (with interesting new research results), without requiring prior knowledge. This is a problem of the second type, in Poincaré's classification of problems.

We now turn to the classification of the graphs of Morse functions on the torus T^2.

Definition. A properly ordered graph is a graph with T triple vertices and K end vertices, which are each given values in such a way that no triple vertex has a value either smaller than all three neighboring vertices, or greater than all three neighboring vertices. We call these values, that order the graph, the "altitudes" of vertices.

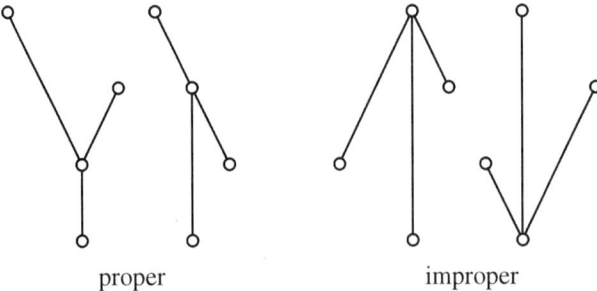

proper   improper

Figure 18. Proper and improper orderings of the vertices of the graph Y.

The exact values for the altitudes are not included in the definition of an ordered graph: two choices of altitudes for the vertices are considered to give the same ordering if a given vertex is lower than a second given vertex in both choices.

Theorem 5. *The number of topologically distinct properly ordered graphs with one cycle and with $T = 4$ triple vertices ($K = 4$ endpoints and $P = 8$ edges) is equal to 550.*

Proof. Such a graph has only one cycle, consisting of 2, 3, or 4 edges. The branch points of the graph form one of the 11 configurations

$A - K$ shown in Figure 19. Here, and below, we show the branch points as squares, and the endpoints (not shown in Figure 19) as small circles.

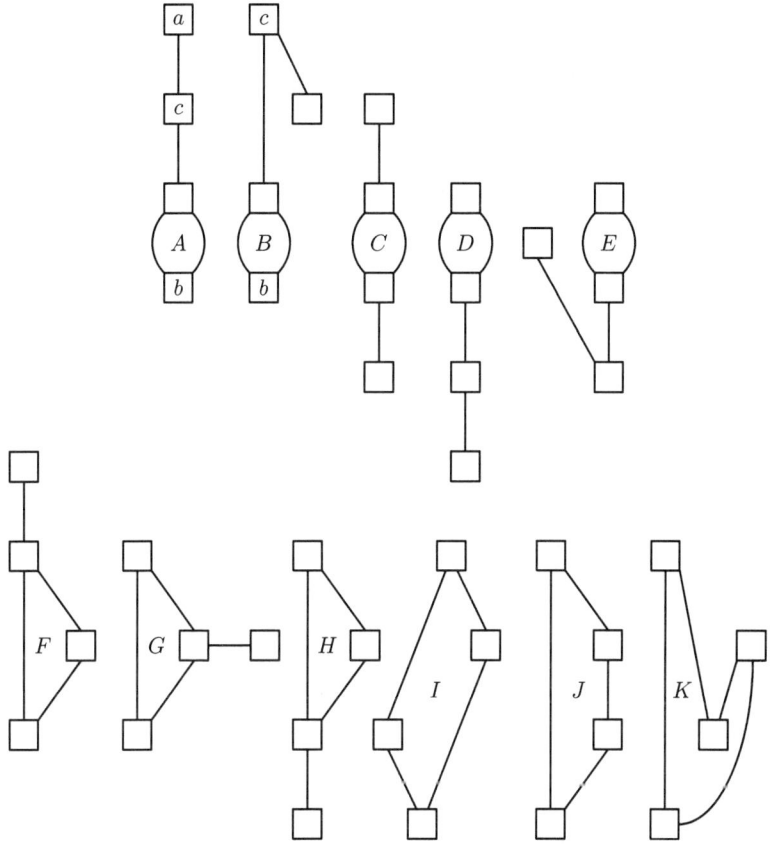

Figure 19. Shortened graphs with one cycle and 4 triple vertices.

Here the altitudes are given by the ordinates of the vertices, except for case G, in which the ordinates of two vertices are chosen to be the same.

We must now compute the number of properly ordered graphs for each of the 11 types. From symmetry (with respect to a horizontal axis), it is clear that these numbers satisfy the relationships:

$$|A| = |D|, \quad |B| = |E|, \quad |F| = |H|,$$

so that in fact there are only 8 essentially different cases: A, B, C, F, G, I, J, K. We now consider these cases.

Case A. There must be a (maximal) edge (aa') leaving the upper branch point at level a upward. There must be a (minimal) edge (bb') leaving the lower branch point at level b downward.

There remain two edges $(a\alpha)$ and $(c\beta)$, leaving the branch points at level a and at level c (respectively).

The graph is defined by the altitudes of the vertices α and β. Four values at the branch points and the value b' divide the semi-axis of values less than a', into 6 parts, so there are 6 (topologically distinct) possibilities for the value of α.

After the values α are chosen, we look at choices for the value of β. The axis of values is divided into 8 parts by the four values at the branch points and the values of (a', b', α) which we have already chosen.

In total, we have 48 (topologically distinct) cases, so *the number of non-equivalent properly ordered graphs of type A is equal to 48.*

Remark. For the further analysis of trigonometric polynomials it is useful to single out those cases in which the number of vertices of the graph with a given altitude is not greater than 2. There are only three such cases (α_1, α_2, and α_3 of the six in Figure 20).

Given any choice of altitude α_1, there are two possibilities for a choice β (larger or smaller than the altitude of c). Given any choice of altitude α_2 there are only two possibilities (both smaller than α_2). Given any choice for α_3 there are no possibilities at all for altitude β. So only 4 cases (shown above) of the 48 properly ordered graphs of type A satisfy the condition (that there are no more than two points of the graph at each altitude).

Case B. The value of altitude α at the four branch points can be found in the four intervals named $\alpha_1, \alpha_2, \alpha_3, \alpha_4$ in Figure 21.

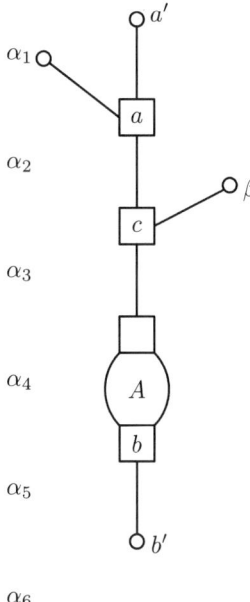

Figure 20. Different charging of a shortened graph of type A.

Once a value of α is chosen, then in order to define a properly ordered graph it remains to choose values for either neighboring vertex of the branch points with altitude α.

Let us call the smaller of these two values β (it must be smaller than the value α_k we have chosen for α). For the choice of a second value, γ, there remain as many intervals as there are intervals into which the values already chosen have divided the interval $\gamma > \beta$. In the case of a choice of value for α of type α_1 we get the following numbers N of intervals:

the choice of k in $\beta_{1,k}$	1	2	3	4
N	4	5	6	7

In just the same way a choice of a value of α of type α_2 leads to intervals $\beta_{2,k}$ (for $k = 1, 2, 3$), with a number N of intervals ($N = 5, 6, 7$) for the placement of values of γ.

In the case of values of α of type α_3 we obtain two intervals $\beta_{3,k}$ $(k = 1, 2)$ with numbers $(N = 6, 7)$ of intervals for the placement of values of γ.

In the case of values of α of type α_4 we obtain a single interval $\beta_{4,1}$ (situated at a lower level than α_4), which is divided into $N = 7$ parts.

Summing up all these cases *we find* $(4 + 5 + 6 + 7) + (5 + 6 + 7) + (6 + 7) + 7 = 22 + 18 + 13 + 7 = 60$ *properly ordered graphs of type B.*

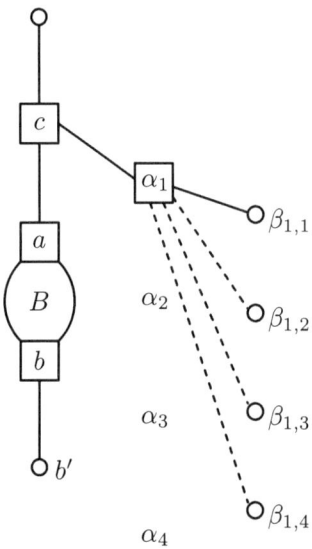

Figure 21. Different chargings of a shortened graph of type B.

Remark. Each of these cases involves a horizontal level with more than two points of the graph. The reason is that from any vertex with value $\alpha_,$, either two edges go upward, or two edges go downwards, and they intersect a horizontal level, while part (bc) of the graph provides a third point on the same horizontal level.

Case C. Let us denote the values at the uppermost and lowermost branch points as a and c. There must be endpoints of edges higher than level a and issuing from a. Let us call the largest of the values at these endpoints a'. Analogously, let us denote by c' the smallest of the values at endpoints of edges issuing from a vertex at level c.

The ordering is determined by the placement of the value α (at the second vertex neighboring the vertex at level a) and the value γ (at the second vertex neighboring the vertex at level c). See Figure 22.

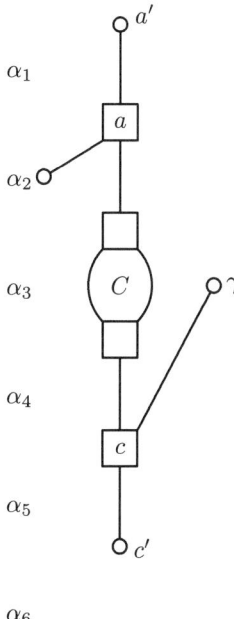

Figure 22. Different charging of a shortened graph of type C.

For a given value of the altitude α the axis of values (smaller than a') is subdivided into six parts (α_k) by the four values at the branch points and the value c'. For a given value of γ the axis of values (greater than c') is subdivided into 7 intervals, by the four values at the branch points and the chosen values of a' and α_k, if $k \leq 5$. (But in the case when $k = 6$ there are only 6 intervals, since $\alpha_6 < c'$).

Thus the total number of properly ordered graphs of type C is $5 \cdot 7 + 6 = 41$.

Remark. Horizontal layers with more than two branch points are missing only in the cases $k = 1$ and 2 (in the notation above): otherwise the edge $(\alpha_k a)$ and the cycle would intersect one horizontal line three times.

For each choice of the value γ there are also only 2 options (so that edge (γc) lies below the cycle). Thus *of the 41 proper graphs of type C only 4 proper graphs satisfy the condition that no three points of the graph lie on the same horizontal levels* as described above.

Case F. I leave as an exercise the enumeration of properly ordered graphs of type F (there are $6 \cdot 8 = 48$ of them). None of these graphs satisfies the condition that there are no three points of the graph on the same horizontal level.

Case G. There are $2(7) + (7 + 6) + (7 + 6 + 5)$ of this type of properly ordered graphs. None of these graphs satisfies the condition that there are no three points of the graph with the same altitude. The proofs of these facts are left as an exercise for the audience.

Cases I, J, K. The number of properly ordered graphs of these types are, respectively, $7 \cdot 8 = 56$, 56, $3 \cdot 3 = 9$. None of these ordered graphs satisfies the condition that no three points lie on the same horizontal level (the proof is left as an exercise).

Conclusion of the proof of Theorem 5. Adding the numbers of properly ordered graphs of the various types enumerated above:

$$|A| = 48, \quad |B| = 60, \quad |C| = 41, \quad |D| = 48, \quad |E| = 60,$$
$$|F| = 48, \quad |G| = 76, \quad |H| = 48, \quad |I| = 56, \quad |J| = 56, \quad |K| = 9,$$

we get the total number, including cases which are symmetric,

$$2(48 + 60) + 41 + 2(48 + 56) + 76 + 9 = 216 + 41 + 208 + 85$$
$$= 257 + 293 = 550. \quad \square$$

Remark 1. I went over and over this boring enumeration trying various ways to count them, until these various ways gave the same result. After several false enumerations I reached a confirmation of the correctness of the final result as well as everything in between.

But I didn't begin the analogous enumeration, say, for $T = 5$ or $T = 6$, although this would be very useful for the discovery of general conjectures (for example, about the growth of the number of properly ordered graphs with a given number of cycles (here this is 1) as the number T of triple vertices grows).

Remark 2. The number of properly ordered graphs of all types, satisfying the condition that no three points of the graph are on the same horizontal level is

$$4(A) + 0(B) + 4(D) + (0(E) + \cdots + 0(K)) = 12.$$

I believe that all these 12 graphs are realized by the trigonometric polynomials

$$f(x, y) = A \sin x + B \sin y + c \sin (x + y) + D \cos (x + y),$$

but I haven't verified this, although one can hope to realize all 12 cases even in a neighborhood of the point

$$A = 1, \quad B = 1, \quad C = -1, \quad D = 0$$

of the indicated four-dimensional space of trigonometric polynomials.[6]

The bifurcation diagram formed by the values $A, B, C, D \in \mathbb{R}^4$, corresponding to the functions with a degenerate (not Morse) critical point or with multiple critical values (assumed at several points) clearly deserve topological and algebraic study (perhaps with the aid of a computer, although it can also be done by hand). This three-dimensional hypersurface in $\mathbb{R}^4$ has the form of a cone erected on its cross section by a three-dimensional plane (for example, the hyperplane $C = 1$), so that this problem is really about the drawing of two-dimensional (algebraic) surfaces in ordinary three-dimensional space.

In doing computer experiments, I found that I make about three times fewer mistakes doing them by hand than doing by computer (for example, even in a simple multiplication of 40-digit numbers). In fact, while unpredictable computer errors are caused by certain

---

[6]Actually, for such trigonometric polynomials the number of saddle points T is less than or equal to 3. Thus the 12 graphs mentioned above are not distinct. The only distinct graphs are the two graphs with $T = 3$ sketched below in Figure 23. (See the table in the end of Section 1.4, where the answers are described in detail.)

cosmic particles, my errors turned out to be always the same, and easily controlled.

Specifically, every time a computation did not fit on one page, I had to copy several (multi-digit) numbers onto the next. It was this copying that was the source of errors: in my small handwriting, the digits 2 and 9 were similar, and so were 3 and 5 — they would be copied incorrectly, and I had to watch carefully (and perhaps others must as well).

4. Algebraic Geometry of Trigonometric Polynomials

To compare the graphs of the special periodic functions

$$f_{A,B,C,D}(x,y) = A\sin x + B\sin y + C\sin(x+y) + D\cos(x+y) \quad (1)$$

with the graphs of general Morse functions (with the same number of critical points) on a two-dimensional torus, I established the following fact, which amazed me.

Theorem 6. *The Morse function (1) on a torus has no more than 8 critical points. A (non-critical) level curve of this function is an elliptic curve (of genus $g = 1$). Its real points on the torus comprise no more than two connected components.*

Actually, there are no more than 6 critical points for the trigonometric polynomial (1). But we won't prove that here.

The ordered graph of a trigonometric polynomial (1) with 6 critical points is one of the two graphs shown in Figure 23 (with the vertices ordered by height).

The remaining 14 ordered graphs of smooth Morse functions on a two-dimensional torus, with 6 critical points, are not represented by trigonometric polynomials (1).

All Morse functions on a torus with identical ordered 6-vertex graphs, and having critical values $\{1, 2, \ldots, 6\}$ can be transformed into each other by diffeomorphisms of the torus.

There exists an infinite set of smooth Morse functions on a two-dimensional torus with 6 critical points and critical values $\{1, 2, \ldots, 6\}$,

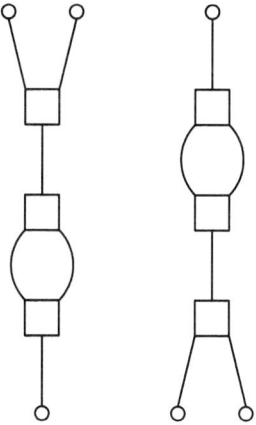

Figure 23

whose ordered graphs are all identical, but not one of them can be transformed into another by a diffeomorphism of the torus homotopic to the identity.

Of these infinitely many Morse functions which are pairwise not reducible to each other, and with ordered graphs shown above, only three functions can be made to coincide with the trigonometric polynomials (1) by a diffeomorphism of the torus which belongs to the connected component of unity in the group of diffeomorphisms of the two-dimensional torus.

A point (in general position) on a cycle of the graph of a function on a two-dimensional torus describes a non-contractible simple closed curve on the torus. This curve, for an appropriate smooth Morse function with 6 critical points, can be any closed curve on the torus (which shows that there are infinitely many Morse functions, which cannot be reduced to each other).

For a trigonometric polynomial (1), a simple closed curve on the torus, as described above, belongs to one of the three types: it can be a parallel, or a meridian, or a diagonal ($x = $ const, $y = $ const, or $x + y = $ const), from which we obtain the three classes of functions mentioned above.

Proof. We use the rational parametrization of the circle. That is, we use ordinary coordinates $t \in \mathbb{R}P^1$ and $\tau \in \mathbb{R}P^1$ on the circles $\{x \mod 2\pi\}$, $\{y \mod 2\pi\}$:

$$\cos x = \frac{1 - t^2}{1 + t^2}, \qquad \sin x = \frac{2t}{1 + t^2},$$

$$\cos y = \frac{1 - \tau^2}{1 + \tau^2}, \qquad \sin y = \frac{2\tau}{1 + \tau^2}.$$

The critical points of the function (1) are defined by the system of equations

$$\begin{cases} A \cos x + C \cos(x + y) - D \sin(x + y) = 0, \\ B \cos y + C \cos(x + y) - D \sin(x + y) = 0. \end{cases}$$

From these two equations we find

$$A \frac{1 - t^2}{1 + t^2} = B \frac{1 - \tau^2}{1 + \tau^2}, \tag{2}$$

from which it follows that

$$\tau^2 = \frac{A(t^2 - 1) + B(t^2 + 1)}{-A(t^2 - 1) + B(t^2 + 1)} = \frac{P_2(t)}{Q_2(t)},$$

where P_2 and Q_2 are polynomials of degree 2,

$$P_2 = (B - A) + t^2(A + B), \quad Q_2(t) = (A + B) + t^2(B - A).$$

We will also need the following formulas, which follow from the previous ones:

$$1 + \tau^2 = \frac{2B(1 + t^2)}{Q_2}, \quad 1 - \tau^2 = \frac{2A(1 - t^2)}{Q_2}.$$

In particular, we have the following amazing fact:

if $t^2 = -1$, then $\tau^2 = -1$ (so that $1 - \tau^2 = 2$).

We have been using just one of the two equations involving the critical points. The second equation

$$A \cos x + C \cos(x + y) - D \sin(x + y) = 0$$

can be written in terms of t and τ in the form

$$A(1 - t^2)(1 + \tau^2) + C[(1 - \tau^2)(1 - t^2) - 4t\tau] + 2D[t(1 - \tau^2) + \tau(1 - t^2)] = 0.$$

In other words, the following quadratic equation holds

$$\tau^2 U + \tau V + W = 0 \tag{3}$$

with coefficients

$$U = (A - C)(1 - t^2) + 2Dt, \quad V = -4tC + 2D(t^2 - 1),$$

$$W - (A + C)(1 - t^2) - 2Dt.$$

Substituting for τ^2 the fraction P_2/Q_2, we get the following solution of equation (3):

$$\tau = \frac{p_4(t)}{q_4(t)}, \quad p_4 = UP_2 + WQ_2, q_4 = -VQ_2. \tag{4}$$

Now the equation $\tau^2 = P_2/Q_2$ takes the form

$$p_4^2 Q_2 = q_4^2 P_2,$$

that is, the form

$$Q_2(p_4^2 - V^2 P_2 Q_2) = 0. \tag{5}$$

This equation has degree 10 in the variable t, so we obtain from it 10 complex critical points (t, τ).

However, two of them are certainly not real: $(i, -i)$ and $(-i, i)$, where

$$t^2 = -1, \tau^2 = -1, t\tau = 1, t + \tau = 0. \tag{6}$$

Indeed, since $t^2 + 1 = 0$ we find

$$P_2 = -24, \quad Q_2 = 2A, \quad V = -4(tC + D),$$

$$U = 2(A - C + tD), \quad W = 2(A + C - tD),$$

so the left hand side of equation (5) acquires the factor

$$[-4A(A - C + tD) + 4A(A + C - tD)]^2 + 16(tC + D)^2 4A^2$$
$$= 16A^2(2C - 2tD)^2 + 64(tC + D)^2 A^2$$
$$= 64A^2(C^2 + t^2D^2 - 2tCD + t^2C^2 + D^2 + 2tCD)$$
$$= 64A^2(C^2 + D^2)(1 + t^2),$$

which vanishes when $1 + t^2 = 0$.

Thus *equation (5) has no more than 8 real roots t, which give us, by equation (4), no more than 8 real critical points of function (1) on the torus.* $\square$

To study the complex surface $\{(x, y) : f(x, y) = c\}$, we use the (affine) coordinates t and τ on complex projective lines, which are the factors of the product $\mathbb{C}P^1 \times \mathbb{C}P^1$ (the complexification of the original torus).

The equation of the surface has the form

$$A\frac{2t}{1+t^2} + B\frac{2\tau}{1+\tau^2} + C\left(\frac{2t}{1+t^2}\frac{1-\tau^2}{1+\tau^2} + \frac{2\tau}{1+\tau^2}\frac{1-t^2}{1+t^2}\right)$$
$$+ D\left(\frac{1-t^2}{1+t^2}\frac{1-\tau^2}{1+\tau^2} - \frac{4t\tau}{(1+t^2)(1+\tau^2)}\right) = c.$$

In other words, the equation of the level curve has the form

$$A[2t(1+\tau^2)] + B[2\tau(1+t^2)] + C[2t(1-\tau^2) + 2\tau(1-t^2)]x$$
$$+D[(1-t^2)(1-\tau^2) - 4t\tau] = c(1+t^2)(1+\tau^2). \qquad (7)$$

For a fixed value of t, this is a quadratic equation in τ. Therefore the image of a complex level surface $\{f(x, y) = c\}$ under the projection

$$\pi : \mathbb{C}P^1 \times \mathbb{C}P^1 \to \mathbb{C}P^1,$$

defined by the formula $\pi(t, \tau) = t$, gives us a (branched) double covering of the Riemann sphere $\mathbb{C}P^1$ with the affine coordinate t.

The branch points t are determined by the condition $\Delta(t) = 0$, where Δ is the discriminant of the quadratic equation (7) with respect to the unknown τ.

It is clear from formula (7) that this discriminant is of degree 4 in the variable t. For a generic level surface of a generic trigonometric polynomial (1) the discriminant will have 4 distinct roots (and will never have more).

A double covering of the sphere with four branch points covers the sphere with the surface of a torus. This can be shown, for example, by computing the Euler characteristic, or using an "Italianate" argument assuming that the covering is given by the formula

$$w^2 = (z^2 - 1)^2 - \varepsilon \text{ (where } \pi(z, w) = z).$$

When $\varepsilon = 0$ the projected complex curve (Figure 24) becomes two Riemann spheres ($w = z^2 - 1$ and $w = 1 - z^2$), intersecting (transversely) at two points ($z = \pm 1, w = 0$).

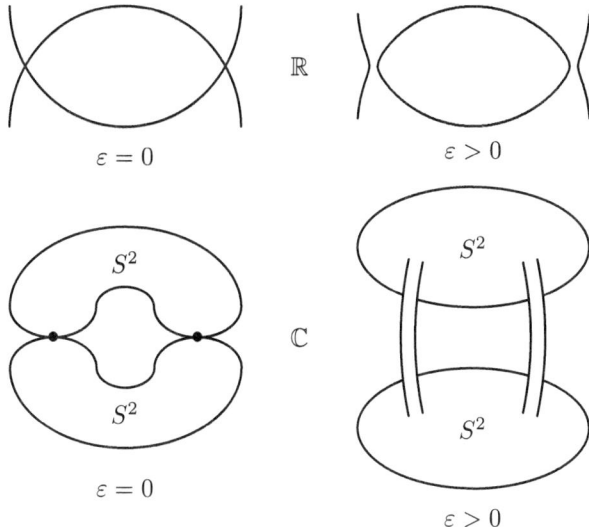

Figure 24. Italian method of studying double covering of the sphere with four branch points.

Taking $\varepsilon \neq 0$ changes a pair of points of intersection into a pair of pipes connecting the two sphere, and the pair of spheres becomes a torus (a surface of genus $g = 1$).

We arrive at the same result if we look at the elliptic curve (Figure 25) given by the equation

$$w^2 = z(z-1)(z-2)(z-3),$$

and its projection on the z−axis parallel to the w-axis (which is a double covering, branched at points $z = 0$, 1, 2, and 3). Cutting the plane of the variable z along the intervals $[0,1]$ and $[2,3]$, we get the sheets w_1 and w_2 of the covering. Pasting along the cuts as shown gives us a torus.

It follows from all this that a (non-singular) level curve of function (1) is an elliptic curve (whose Riemann surface is a torus of genus $g = 1$).

But a real elliptic curve cannot have more than $g + 1 = 2$ connected components (by Harnack's theorem, given above in Section 1.2).

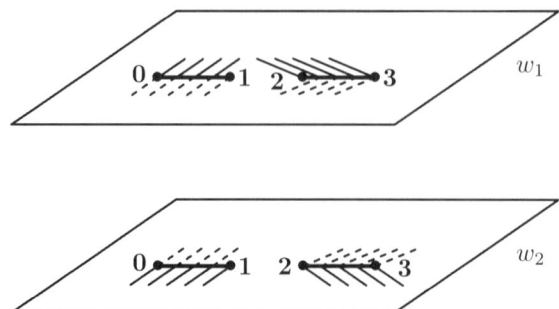

Figure 25. The construction of the Riemann surface of an elliptic curve by pasting together two Riemann spheres with a pair of cuts on each of them.

Therefore a real level curve of the trigonometric polynomial (1) cannot have more than two connected components on the torus

$$\{x\,(\mathrm{mod}2\pi),\ y\,(\mathrm{mod}2\pi)\}.$$

It follows from this that in the corresponding graph there cannot be more than two points on one horizontal level.

From what we have proved in Section 1.3, out of the 550 properly ordered graphs of Morse functions on a torus with $T = 4$ saddle points, only 12 graphs have no more than two points on each horizontal level.

This means that the graphs of trigonometric polynomials of the form (1) realize at most 12 of the 550 topologically possible graphs with $T = 4$. I believe that all 12 of these graphs are realized by trigonometric polynomials of the form (1), and the reason for talking about trigonometric polynomials (in this lecture about Hilbert's 16th problem dedicated to the real algebraic geometry of ordinary polynomials) is that I myself could not realize these graphs.

Neither experts in computational mathematics with their computers, nor algebraic geometers with their axiomatizations, have made any essential contribution to the solution of these real (i.e., related to $\mathbb{R}$) problems. The greatest achievements in this area, started by Descartes and Newton, Hurwitz and Klein, Harnack and Hilbert, belong to the Russian school of mathematics: I. G. Petrovsky, and V.

A. Rokhlin, O. Ya. Vrio and V. M. Kharlamov, G. M. Polotovsky and E. I. Shustin.

But, unfortunately, these same natural questions about the topological structure of ordinary and trigonometric polynomials remain open, it seems, through all these investigations.

For example, let us look at the topological classification of real polynomials in one variable of degree $n + 1$ with n real critical points with different critical values, and with the leading term x^{n+1}.

The numbers N of topological types for small values of n is not hard to compute:

n	0	1	2	3	4	5	6	7	8
N	1	1	1	2	5	16	61	272	1385

(we consider the "topological type" up to topological transformations preserving the orientation of the coordinate axes). This remarkable sequence of numbers $N(n)$, easily recognizable by Euler's number 61, delivers the Taylor expansion of secant and tangent:

$$\sum_{n=0}^{\infty} \frac{N(n)t^n}{n!} = \sec t + \tan t; \quad \sec t = 1 + \frac{1}{2!}t^2 + \frac{5}{4!}t^4 + \frac{61}{6!}t^6 + \cdots,$$

$$\tan t = \frac{1}{1!}t + \frac{2}{3!}t^3 + \frac{16}{5!}t^5 + \frac{272}{7!}t^7 + \cdots.$$

The theorems of the present lecture arose from an attempt to expand these results to functions of several variables.

The question of the topological classification of real polynomials was not included by Hilbert in his classical formulation of the problem, but these dogmatic formulations hypnotized researchers who were trying to solve the classical problem, rather than get to the heart of the matter.

At the International Mathematical Congress in Stockholm in 1962 I talked about my solution of Birkhoff's stability problem for non-resonant equilibrium states of Hamiltonian systems. Having solved this classical problem, I didn't notice that I had proved even more, namely, the stability of the majority of the resonant cases (for resonances of order 5 and higher), because the classical problem excluded all resonances. The American mathematician Y. Moser immediately noticed the property that in fact I had proved (the stability of "weak resonances"), but the classic formulation of the classical problem hindered me from including my own result in the report to the Congress.

Happily, neither the presentation at the congress, nor the elections to the Academies, nor the inclusion of the problem in Hilbert's list, nor awards like the Fields Medal or the Nobel Prize, have had any great influence on the development of mathematics, nor even on other sciences. So I hope that the Dubna summer school will have a greater influence (and will result in the solution by my listeners of at least some of the problems discussed in the present lecture).

For example, in 2006 L. Nicolaescu, continuing the research described in a Dubna lecture of 2005, proved a conjecture stated in that lecture about the growth rate of order of T^{2T} of the number of types of Morse functions with T saddle points on S^2 (see his article [16]). It is not clear how the number of types of polynomials of degree n grows. (Perhaps it grows as a power of n, and not as an function with n in the exponent.)

The work of Nicolaescu reveals an amazing connection between the problem of classifying the topological types of Morse functions on S^2 and the theory of mirror symmetry in quantum field theory (constructed by A. B. Giventhal), although this is not clearly stated in Nicolaescu's article.

The point is that he found a recurrence relation between the numbers $\psi(T)$ of types of Morse functions on the sphere with T saddle points, having proved that these numbers of topological types (17746 and so on) appear in the investigation of the expansion of several elliptic integrals as series in the powers of the parameter, on which the integrand $(x^3 + ax + b)^{-\frac{1}{2}}$ depend (rationally).

Proofs of Giventhal's theorems about mirror symmetry are also based on the expression of integer characteristics of Hodge numbers (of complex algebraic manifolds of complex dimension 3) in terms of the coefficients of the series expansion of several special Abelian integrals (along cycles of the "mirror image" of the manifold).

The same topological characteristic of the manifold-image can be expressed in the same way in terms of the analogous integrals along cycles of the original three-dimensional manifold. It is interesting that the two-dimensional analog to mirror symmetry in physics turned out to be the "strange parity" of triangles on the Lobachevskian plane that had been discovered earlier (Arnold, 1974).

What the mirror image of the problem for the topological classification of smooth Morse functions on the sphere S^2 might be remains, unfortunately, unclear.

The topological classification of trigonometric polynomials and functions on the torus, begun above in section 3, led me to unexpected results, which will be described in a lecture in Dubna in 2006. There are finitely many classes in several classification problems, but on others there are infinitely many. The answers to the question of classifying trigonometric polynomials of the form (1) and smooth functions on the torus with six fixed critical values are as shown in the table below.

C^∞ analysis: smooth Morse functions with 6 critical points on the torus T^2	16 classes	∞ classes
Algebra: trigonometric polynomials of the form (1)	2 classes of polynomials	6 classes of polynomials
Classification with respect to	Diff(T^2) classification of functions up to diffeomorphisms of the torus	Diff$_0(T^2)$ classification of functions up to diffeomorphisms of the torus T^2, homotopic to the identity.

Editor's notes

1. The topological classification of curves of degree 6 exists not only for curves with 11 ovals, but for all non-singular curves of degree 6. Namely, any such curve, with one exception, is homeomorphic

to a union of ovals obtained from the curves shown in Figure 3 by removing any set of ovals. The exception is a nest of three ovals (for example, the union of three concentric circles). (See the survey article [**15***] by V. Kharlamov and O. Viro, available on the web.)

2. Gudkov's congruence has a more elementary statement. Consider a curve of degree $2k$ with the maximal possible number of ovals. An oval of this curve is called even (odd), if it is contained in an even (odd) number of other ovals. Let p and q be the number of even and odd ovals respectively. (For example, for the curves in Figure 3, (p, q) is $(10, 1)$, $(2, 9)$, and $(6, 5)$ respectively.) The congruence is $p - q \equiv k^2$ mod 8. (See again the article of Kharlamov and Viro mentioned in Editor's note 1.)

3. A smooth (infinitely differentiable) function f of n variables $x_1, \ldots, x_n$ is called a *Morse function*, if every critical point (that is a point where all partial derivatives $\partial f / \partial x_i$ are zeroes) is *nondegenerate*; that is, the Hessian

$$\begin{vmatrix} \dfrac{\partial^2 f}{\partial x_1^2} & \cdots & \dfrac{\partial^2 f}{\partial x_1 \partial x_n} \\ \vdots & \ddots & \vdots \\ \dfrac{\partial^2 f}{\partial x_n \partial x_1} & \cdots & \dfrac{\partial^2 f}{\partial x_n^2} \end{vmatrix}$$

is different from zero at this point. For a Morse function of two variables, all critical point are local maxima, local minima or saddle points.

Almost all smooth functions are Morse functions in the sense that every smooth function can be approximated (with all its derivatives) by Morse functions, and a function sufficiently close (with all derivatives) to a given Morse function is a Morse function. Less formally speaking, a randomly chosen smooth function is a Morse function.

It is not hard to prove that a Morse function has finitely many critical points on every compact domain.

4. Let f be a Morse function on the sphere (or other compact manifold). We collapse every component of every level hypersurface to

a point. The resulting space is a graph. Its vertices correspond to the critical points of the function, and two vertices are connected with an edge if the two corresponding critical points may be connected with a path not crossing components of level hypersurfaces containing other critical points.

Since the function f is constant on every set which is collapsed to a point of the graph, we may consider the function f as defined on the graph.

5. If f is a polynomial of degree n, then the partial derivatives $\partial f/\partial x$ and $\partial f/\partial y$ are polynomials of degree $n-1$, so the system of equation $\partial f/\partial x = 0$, $\partial f/\partial y = 0$, if it is not degenerate, has $(n-1)^2$ solutions, plus one more for the point at infinity.

Note also that if the degree of the polynomial f is odd, then the value of $f(P)$ may approach both $+\infty$ and $-\infty$ as P approaches ∞, so in this case f does not give rise to any function on the sphere. This is why we restrict ourselves to considering polynomials of even degree $n = 2k$.

6. The smaller graph has two fewer vertices than the original graph; that is, it has $2T$ vertices. This smaller graph is again a tree, and the number of edges of a tree is always one less than the number of vertices. Thus, the smaller graph has $2T - 1$ edges.

Lecture 2

Combinatorial Complexity and Randomness

> "To err is human, but to really foul things up requires a computer."
>
> *from "Murphy's Laws"*

> "Faith and knowledge are related as two scales of a balance: as one rises the other falls."
>
> *A. Schopenhauer*

Today's lecture fits badly into the traditional divisions of mathematics into domains (such as "number theory" or "point set topology").

The question of *how random a given function or sequence is*, does not belong to mathematics, although all of us understand, for example, that the sequence

$$001\ 001\ 001\ 001$$

is less random than the sequence

$$010\ 010\ 111\ 001.$$

I will now attempt to give these words some mathematical meaning. In particular, we will associate certain geometric objects with such sequences, and the natural geometric characteristics of these objects will serve as a measure of complexity of the original sequence.

This analysis can be regarded as belonging to the geometry of finite function spaces, to combinatorics, to the arithmetic of finite fields, to the statistics of random processes, to the linear algebra of matrices, to graph theory, to the arithmetic of Jordan normal forms of operators, and to the theory of finite differences and numerical integration, but I prefer not to confine this analysis to any of these narrow domains.

1. Binary Sequences

Let x be a sequence of n zeroes and ones,

$$x = (x_1, x_2, \ldots, x_n), \quad x_j \in \mathbb{Z}_2 \quad (= \mathbb{Z}/2\mathbb{Z}).$$

The set M of all such sequences is finite and consists of 2^n elements. We can think of these elements as the vertices of an n-dimensional cube (Figure 1), but also we can regard $\mathbb{Z}_2^n$ as an n-dimensional vector space (over the field $\mathbb{Z}_2$ consisting of two elements 0 and 1, with the Mickey Mouse operations:

$$1 + 1 = 0 + 0 = 0, \quad 0 + 1 = 1 + 0 = 1,$$

$$0 \cdot 0 = 0 \cdot 1 = 1 \cdot 0 = 0, \quad 1 \cdot 1 = 1).$$

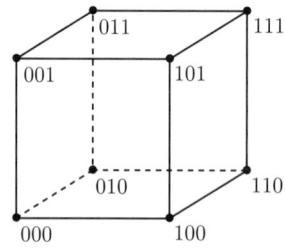

Figure 1. The vertices of an n-dimensional cube for $n = 3$.

To investigate how orderly a sequence x is, we will think of it as a *function with values 0 and 1, defined on a set of n elements j*:

$$x : \{1, 2, \ldots, n\} \to \mathbb{Z}_2$$

(so that $x(j) = x_j$).

In order to study such a function, we will follow Newton's recommendation and consider the sequence of its differences,

$$y_j = x_{j+1} - x_j \in \mathbb{Z}_2.$$

In order that the sequence of differences have the same length as the original sequence of n elements, we define x_{n+1} to be x_1, turning our sequence of the length n into a loop. (Or, we can say we consider the function x of the natural argument j periodic, with period n.) In this sense we can think of the domain of the function x as a finite circle of n points,

$$\mathbb{Z}_n = \mathbb{Z}/n\mathbb{Z}.$$

This function

$$x : \mathbb{Z}_n \to \mathbb{Z}_2$$

is not at all assumed to be linear.

The simplest functions among these are the constants, $x = 0$ and $x = 1$. We will consider polynomials of a lower degree as simpler functions than polynomials of higher degree. The definitions that follow will give this idea a precise meaning.

According to Newton, for a polynomial of degree less than m (and only for them) the operation of taking differences leads to zero after it is applied m times:

$$A^m x = 0.$$

The operator of taking differences $A : \mathbb{Z}_2^n \to \mathbb{Z}_2^n$ maps the finite set M of sequences onto itself. Therefore we can associate with it the following remarkable *graph of the operation*, which has $|M| = 2^n$ vertices x: *exactly one arrow leads out from each vertex x, and it leads from the vertex x to the vertex Ax.*

For example, let us construct this graph for the case $n = 3$, with 8 vertices. By definition, we find

$$A(0,0,0) = (0,0,0), \quad A(0,0,1) = (0,1,1), \quad A(0,1,0) = (1,1,0),$$

$$A(0,1,1) = (1,0,1), \quad A(1,0,0) = (1,0,1), \quad A(1,0,1) = (1,1,0),$$

$$A(1,1,0) = (0,1,1), \quad A(1,1,1) = (0,0,0).$$

In order to avoid writing such long formulas, we have another notation for $x = (x_1, \ldots, x_n)$. We will use integers (or integers modulo 2^n) written as binary numerals (in the base-two system, as in a computer):

$$X = x_1 2^{n-1} + x_2 2^{n-2} + \cdots + x_n 2^0.$$

In this notation, the vertex $x = (1,1,1)$ of our cube becomes the number $X = 7$. The entire table given above for the action of the operation A of taking differences of binary sequences with period $n = 3$ takes the form of an oriented graph with eight vertices:

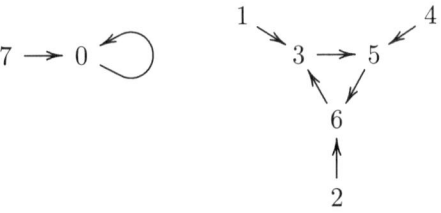

In the graph of any transformation of a finite set onto itself, only one arrow leads out from each vertex. It is easy to prove

Theorem 1. *Every connected component of a graph of any transformation of a finite set onto itself contains one and only one cycle.*

The entire connected component can be obtained from this attractor cycle by appending to each vertex of the cycle the tree of vertices attracted to that vertex.

For example, for the operation of taking differences in $\mathbb{Z}_2^3$ we found two components, with cycles of length 1 and 3, furnished with the simplest of trees with two vertices

Proof of Theorem 1. We look at the orbit of any point x under the transformation A (this orbit consists of the points $x, Ax, A^2 x, \ldots$).

Since the whole set on which the transformation A acts is finite, the points in this sequence cannot all be distinct: $A^p x = A^q x$ for some (unequal) p and q. We take the first such pair. For example, suppose $p > q$. Then $A^r y = y$ for $r = p - q$, $y = A^q x$; that is, we have found a cycle of period r (in the connected component of any point x; that is, in any component of the graph).

If there were two cycles in some component connected by a finite chain of edges of the graph, $(a, b, \ldots, z)$, then edge a would be directed towards the first cycle, and edge z towards the second one. So this chain would contain a vertex from which edges started directed to both cycles. This is not possible, since only one arrow can lead out of each vertex (and an edge leading out of a vertex in a cycle must belong to that cycle).

Thus there is only one cycle in a component, and all the other edges in this component lead to this cycle. This completes the proof of Theorem 1. $\qquad\square$

Let us use the following notation: O_m denotes a cycle of m edges. If T is a tree, then $O_m * T$ will denote a graph consisting of m copies of the tree T, with the edges of each copy directed towards the root of the tree, and the roots of the trees forming a cycle O_m in the graph.

In this notation, the previous description of the operation of taking differences $A : \mathbb{Z}_2^3 \to \mathbb{Z}_2^3$ takes the form of a graph with two components, $(O_1 * T_2)$ and $(O_3 * T_2)$.

The symbol T_{2^n} will denote the binary tree with n levels above the root and 2^n vertices, whose edges are directed towards the root (Figure 2).

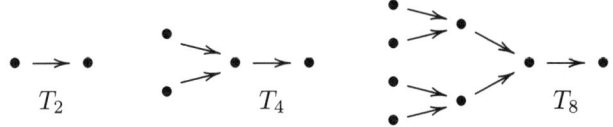

Figure 2. Binary root trees.

By definition, a tree is binary if exactly two edges point at each vertex of each level, except the first and last level. Each edge starts at next level. Also, there is only one edge pointing at the root of the tree on the first level, and this edge starts at the (unique) vertex of the second level. For example, the symbol $(O_2 * T_4)$ denotes the directed graph with eight vertices shown in Figure 3.

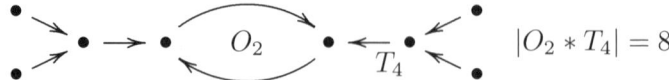

Figure 3. The forest of trees T_4 on the cycle O_2.

Our plan for investigating the complexity of a sequence $x \in \mathbb{Z}_2^n$ consists in looking at point x as a vertex of the graph of Newton's operation (taking differences) $A : \mathbb{Z}_2^n \to \mathbb{Z}_2^n$.

We will consider the sequence x to be more complex if the component of the graph of the operator A that x belongs to is larger (that is, the cycle of this component is longer, and the forest of trees attached to this cycle is bigger). For cycles of equal lengths, we will take into account how far the point x is from the cycle (that is, the height of the branch of the tree where x is located).

To do this, we must first list all the components of the graph, their cycles, and their forests. This is not such a simple task, and I know the whole answer only for $n \leq 12$.

2. Graph of the Operation of Taking Differences

Just as we have done above for periodic sequences of length $n = 3$, I have counted the graphs of the operation of taking differences of periodic binary sequences of period n, $A : \mathbb{Z}_2^n \to \mathbb{Z}_2^n$ for all $n \leq 12$.

For larger values of n I did not do this, because for $n = 12$ the number 2^n of vertices of these graphs is just 4096, which allows one to draw the whole graph on a single page. But for larger values of n the graph takes up several pages and I got lost in my own computations, done without computer.

And yet a continuation of my count for larger values of n would be interesting, in particular because the analysis for $n \leq 12$ leads to a whole range of fascinating hypotheses, which would be nice to confirm experimentally before we try to prove them.

It is exactly this possibility of quickly continuing my investigations (without knowing anything in advance) that made me decide to include this subject in the present series of lectures for high school students in Dubna. I hope that my listeners will achieve new success.

Theorem 2. *The graphs of the operation A of taking differences of binary periodic sequences of length $n \leq 12$ have the structures listed in the table below.*

n	b	components of the graph of the operation A	relation
2	1	$(O_1 * T_4)$	$A^2 = 0$
3	2	$(O_3 * T_2) + (O_1 * T_2)$	$A^4 = A$
4	1	$(O_1 * T_{16})$	$A^4 = 0$
5	2	$(O_{15} * T_2) + (O_1 * T_2)$	$A^{16} = A$
6	4	$2(O_6 * T_4) + (O_3 * T_4) + (O_1 * T_4)$	$A^8 = A^2$
7	10	$9(O_7 * T_2) + (O_1 * T_2)$	$A^8 = A$
8	1	$(O_1 * T_{256})$	$A^8 = 0$
9	6	$4(O_{63} * T_2) + (O_3 * T_2) + (O_1 * T_2)$	$A^{64} = A$
10	10	$8(O_{30} * T_4) + (O_{15} * T_4) + (O_1 * T_4)$	$A^{32} = A^2$
11	4	$3(O_{341} * T_2) + (O_1 * T_2)$	$A^{342} = A$
12	24	$20(O_{12} * T_{16}) + 2(O_6 * T_{16}) + (O_3 * T_{16})$ $+ (O_1 * T_{16})$	$A^{16} = A^4$

Column b (the "Betti numbers") lists the number of connected components of the graph. The components themselves are listed in the next column. The notation $20(O_{12} * T_{16})$ in the row for $n = 12$ means that there are 20 components (isomorphic to each other) with cycles of period 12, with a (four-level) binary tree with 16 branches attached at each point of the cycle of the graph.

Thus, for $n = 12$ the 24 components of the graph give use the following number of vertices:

$$20 \cdot 12 \cdot 16 + 2 \cdot 6 \cdot 16 + 1 \cdot 3 \cdot 16 + 1 \cdot 1 \cdot 16 = (240 + 12 + 3 + 1)16$$
$$= 256 \cdot 16 = 2^{12},$$

(as must happen since $|\mathbb{Z}_2^{12}| = 2^{12}$). The total number of vertices of all its cycles is $20 \cdot 12 + 2 \cdot 6 + 1 \cdot 3 + 1 \cdot 1 = 256$.

The last column of the table lists the identities which are satisfied by the linear operators A. For example, for the components $(O_3 * T_2)$ and $n = 3$ we note that the point $y = Ax$ always belongs to the cycle, so that $A^3 y = y$, and $A^4 = A$.

However, this identity could have been predicted in advance algebraically, before the graph (whose construction is facilitated by the identity) was drawn.

Let us denote by $\delta : \mathbb{Z}_2^n \to \mathbb{Z}_2^n$ the linear operator which cyclically permutes sequences of length n (so that $(\delta x)_j = x_{j-1}$ for $j \in \mathbb{Z}_n$).

Clearly $A = 1 + \delta$ (since in calculating modulo 2 a difference coincides with a sum), and $\delta^n = 1$ (since δ is a cyclic shift of a regular n-gon through an angle $2\pi/n$).

Now for $n = 3$ we can compute the successive powers of the operator A of taking differences as follows:

$$A = 1 + \delta, \quad A^2 = 1 + 2\delta + \delta^2 = 1 + \delta^2,$$

$$A^3 = (1 + \delta)(1 + \delta^2) = 1 + \delta + \delta^2 + \delta^3 = \delta + \delta^2$$

(since $\delta^3 = 1, 1 + 1 = 0$). Finally,

$$A^4 = (1 + \delta)(\delta + \delta^2) = \delta + \delta^2 + \delta^3 = \delta + 1 = A.$$

Using computations such as these, we can prove the remaining relations in the last column of the table.

Remark. Looking at the table, we can make a series of interesting observations, some of which have already been proven as theorems. For example, if $n = 2^k$, there is only one component of the graph, of period 1, since the entire graph becomes a tree T_{2^n} with 2^{2^k} vertices and with root $x = 0$.

In this case the whole function space of n-periodic functions with values in $\mathbb{Z}_2$ coincides with the ring of "polynomials". (In the general case, the subring of "polynomials" forms the last connected component drawn, $O_1 * T_r$, where $r = 2^{2^k}$ for $n = 2^k(2l + 1)$.)

By a "polynomial" I mean here a polynomial with rational coefficients and integer values

$$x(j) = a_0 j^m + a_1 j^{m-1} + \cdots + a_m$$

at integer points j, taken modulo 2, so that the values end up in $\mathbb{Z}_2$.

Examples of such "polynomials" are provided, miraculously, by the binomial coefficients:

$$\binom{j}{2} = \frac{j(j-1)}{2}, \quad \binom{j}{3} = \frac{j(j-1)(j-2)}{6}, \quad \text{etc.}$$

The "polynomials" of period n form a subring of the ring of n-periodic functions (with values in $\mathbb{Z}_2$).

For further work, an interesting question is what the period is of the "polynomial" $x(j) = \binom{j}{k}$ mod 2 for fixed k, and also the dimension of the vector space of "polynomials" of period n with values in $\mathbb{Z}_2$ (over the field $\mathbb{Z}_2$): how many such "polynomials" are linearly independent?

These questions about the arithmetic of binomial coefficients modulo 2 are easy to answer using the geometry of Pascal's triangle, and I leave these to the readers, in the hope that they have seen this triangle someplace (although it does not appear in contemporary textbooks of probability theory for pre-college students).

I leave the reader, for the moment, to guess at the enigmatic behavior for varying n of the maximal periods T (and other periods) of cycles of the components of the graph for the operator that takes differences in $\mathbb{Z}_2^n$:

n	2	3	4	5	6	7	8	9	10	11	12
T	1	3	1	15	6	7	1	63	30	341	12

If the period T of the largest cycle differs from 1, then it is divisible by n, and for some reason the quotient has the form $2^s - 1$.

For example, $341 = 11 \cdot 31$.

By the way, 341 is a remarkable number: it yields a first example disproving the false Chinese converse of "Fermat's Little Theorem" which survived for millennia. The ancient Chinese thought that if $2^a \equiv 2 \bmod a$, then the number a is prime. But for $a = 341$ this congruence holds, since

$$2^{11} \equiv 2 \pmod{11}, \quad 2^{11} \equiv 2 \pmod{31},$$

$$2^{31} \equiv 2 \pmod{11}, \quad 2^{31} \equiv 2 \pmod{31};$$

and yet 341 is not prime (see Editor's note 2, page 80).

It is also amazing, as the table shows, that each connected component of the graph has the form $O_m * T_{2^k}$, where the binary tree T_{2^k} is the same as the one that appears in the ring of "polynomials" corresponding to the period.

This fact is explained by the elementary geometry of the linear operator

$$A : \mathbb{Z}_2^n \to \mathbb{Z}_2^n.$$

This geometry asserts that the set $\text{Ker}(A^s)$ of solutions of the x-linear homogeneous equation

$$A^s x = 0$$

is a vector subspace of the vector space $\mathbb{Z}_2^n$, while the set of solutions of a non-homogeneous equation (with right hand side y) is an affine subspace parallel to it.

The subspaces of solutions to homogeneous equations with various values of s form an increasing sequence:

$$\text{Ker}(A) \subseteq \text{Ker}(A^2) \subseteq \text{Ker}(A^3) \subseteq \cdots \subseteq \text{Ker}(A^\infty)$$

where $\text{Ker}(A^\infty)$ simply denotes the largest of the spaces in the sequence. (The notation Ker originates from the word "Kernel", the name for the space of solutions of a homogeneous equation.)

This space $\text{Ker}(A^\infty)$ is also the ring of "polynomials", since by Newton's Theorem, the "polynomials" form the union of the whole sequence (of spaces of polynomials of various degrees). In the graph of the operator A this is the component attracted by the vertex 0 of period 1.

The images of the powers of the operator A form another sequence of subspaces:

$$\mathbb{Z}_2^n \supseteq A(\mathbb{Z}_2^n) \supseteq A^2(\mathbb{Z}_2^n) \supseteq A^3(\mathbb{Z}_2^n) \supseteq \cdots .$$

This sequence of vector subspaces decreases. I will denote by $A^\infty(\mathbb{Z}_2^n)$ the intersection of all these subspaces. (We will not worry here about meaning of the operator A^∞, just as we ignored this question in the case of the space of solutions to the homogeneous equation.)

The space $A^\infty(\mathbb{Z}_2^n)$ can be described simply in terms of the graph of the operator A. It is exactly made up of all points of all the cycles of all components (since any other vertex of the graph, whose forests have height h, has no vertex which is higher than it by more than h, and so does not belong in the image of the operator A^{h+1}).

For example, it follows from all this that the sum Σ of the lengths of all the cycles of the graph is a power of two (namely, 2^q, if the vector space $A^\infty(\mathbb{Z}_2^n)$ has dimension q over $\mathbb{Z}_2$). This can be observed in our table:

n	2	3	4	5	6	7	8	9	10	11	12
Σ	1	4	1	16	16	64	1	128	256	1024	256

All the assertions of Theorem 2 can be verified directly by computation. I will not give these computations, in the hopes that my listeners (armed, possibly, with computers) will do the computations themselves not just for $n \le 12$ but for much larger periods n of periodic binary sequences.

3. Logarithmic Functions and Their Complexity

The table shows that if n is not a power of 2, then not all n-periodic functions with values in $\mathbb{Z}_2$ will be "polynomials". In this case, we can consider these non-polynomial functions x "more complicated than any polynomial". If we wanted, we could continue the hierarchy of functions of our finite function space (which has only 2^n elements) by looking at exponents, sines, quasi-polynomials, and so on, and comparing their "complexity" (depending, for example, on the simplest differential equation they satisfy. (See Editor's note 3, page 107)).

I will not do this now but rather will look at one special function—the discrete "number theoretic logarithm" with values 0 and 1, which somehow turns out to be a "very complicated" function, in our sense of that term.

To define this arithmetic logarithm I assume that $n + 1 = p$ is an odd prime number.

Let a be a primitive root modulo p, so that $a^{p-1} \equiv 1 \bmod p$, and the n remainders

$$\{a, a^2, \ldots, a^{p-1}\}$$

give us all the non-zero remainders k upon division by p.

If

$$a^\ell \equiv k \bmod p,$$

we will call the number ℓ (or rather the remainder when ℓ is divided by $n = p - 1$) the *arithmetic logarithm* of the number (or the remainder) k:

$$\ell = \log_a k.$$

We have defined the function ℓ of the argument k. The function L which we need has values in $\mathbb{Z}_2$: it is the remainder upon division of ℓ by 2:

$$L(k) \equiv \ell(k) \bmod 2.$$

Notice that because the number $p - 1$ is even, the value of $L(k)$ is well-defined (despite the fact that ℓ is defined only as the remainder upon division by $p - 1$).

We will examine these values as a binary sequence of length $p - 1 = n$:

$$L(1), \ L(2), \ldots, L(n).$$

The value at the point k is zero if and only if $k \neq 0$ is a quadratic residue modulo p. It is equal to 1 if $k \neq 0$ is a quadratic non-residue.

It is evident from this that our "arithmetic binary logarithm" $L : \mathbb{Z}_n \to \mathbb{Z}_2$ does not depend on the choice of the primitive root a with which we started our definition, but depends only on the number $n = p - 1$.

Direct computation of the position of this "arithmetic logarithm" in the graphs of Theorem 2 shows that *this special function is either maximally complicated* (among all binary functions of period n) *or almost maximally complicated* (in the sense of complexity that we defined using the geometry of the graph of the operation of taking differences).

These computations, which are not complicated in themselves, are rather long, and to describe their results we will describe a function $L \in \mathbb{Z}_2^n$ by a binary number X with n using the digits $L(k)$:

$$X = L(1)2^{n-1} + L(2)2^{n-2} + \cdots + L(n)2^0 \bmod 2^n.$$

In the simplest case (when $p = 3$, $n = 2$), the logarithmic function L is determined by a geometric progression of n remainders

$$\{2^\ell \bmod 3\} = (2, 1).$$

Thus we have $\ell(2) = 1$, $\ell(1) = 2$, so that

$$L(1) = 0, \quad L(2) = 1, \quad X = 1 \bmod 4.$$

The graph of the operator A in the case $n = 2$ has the form

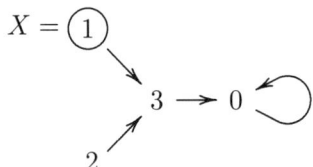

The point $X = 1$ is the most complicated binary function of period 2, since this vertex of the tree T_4 is furthest from the cycle (of the root 0).

For the following simple cases ($p = 5$, 6, 11, 13) the computations are completely analogous, but longer, so I present only their results.

Theorem 3. *In the graphs of the operation of taking differences of binary sequences of period $n = p - 1$, the arithmetic logarithms L correspond to the vertices X given in the following tables for the cases $p = 5$, 7, 11 and 13.*

For $p = 7$, 11 these are the most complex binary functions of period $n = p-1$. For $p = 5$, 13 these are almost the most complicated. (For $p = 7$, 11, vertex X is located at the furthest point from the cycle, and for the other cases it is 1 closer).

The case $p = 5$, $n = 4$. If we take $a = 2$, we find

$$\ell(1) = 0, \quad \ell(2) = 1, \quad \ell(3) = 3, \quad \ell(4) = 2,$$

from which we obtain $X = 6$ for the "arithmetic binary logarithm".

The component of the graph that we need has the form $O_1 * T_{16}$ (see the diagram below).

The case $p = 7$, $n = 6$. If we take $a = 3$, we get $X = 11$ for the "arithmetic binary logarithm".

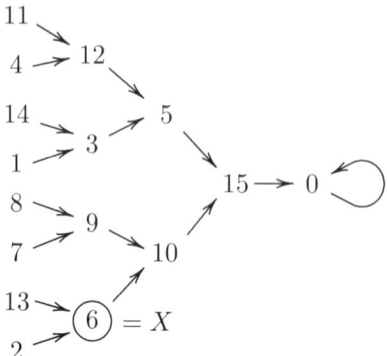

The component of the graph that we need has the form $O_6 * T_4$, and I will give only the orbit of point X in this graph:

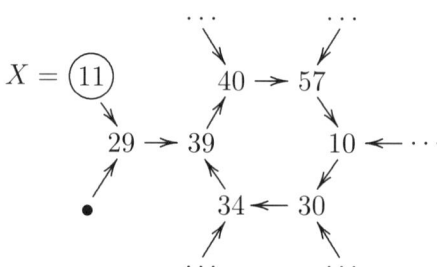

The case $p = 11$, $n = 10$. Again taking $a = 2$ for the initial remainder we obtain the geometric progression $(2, 4, 8, 5, 10, 9, 7, 3, 6, 1)$ (mod 11), from which we find $L = (0, 1, 0, 0, 0, 1, 1, 1, 0, 1)$, and $X = 285$.

The corresponding component has the form $O_{30} * T_4$, and, of its 120 vertices, I will write out only the 32 vertices of the orbit of the binary arithmetic logarithm X:

The case $p = 13$, $n = 12$. The initial remainder $p = 2$ gives us the values

$$\log_2 k = (12, 1, 4, 2, 5, 11, 3, 8, 10, 7, 6)$$

for $k = (1, 2, \ldots, 12)$, from which we obtain $X = 1266$ for the binary arithmetic logarithm.

This vertex belongs to the largest component of the graph, $O_{12} * T_{16}$. Of the 192 vertices of this component only 15 vertices occur in

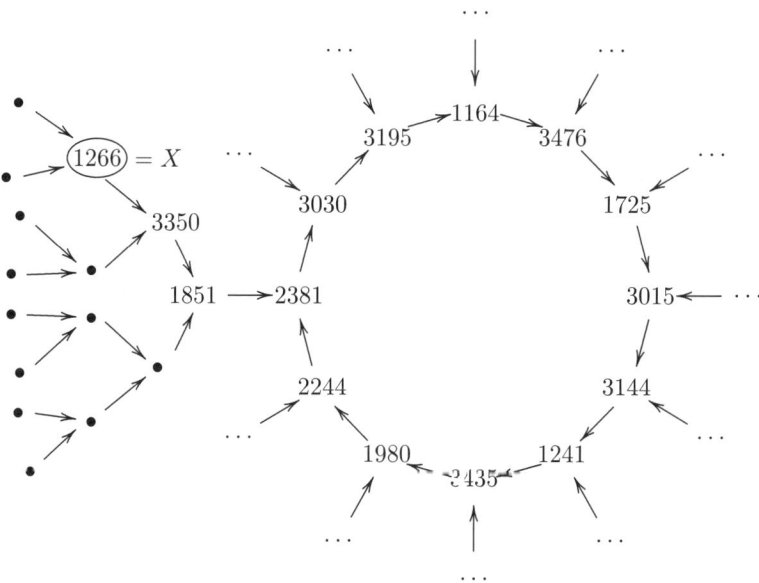

$X = \widehat{285} \twoheadrightarrow 807 \twoheadleftarrow \cdots$

$\downarrow$

$\twoheadrightarrow 216 \twoheadrightarrow 360 \twoheadrightarrow 952 \twoheadrightarrow 201 \twoheadrightarrow 347 \twoheadrightarrow 1005 \twoheadrightarrow 54 \twoheadrightarrow 90 \twoheadrightarrow 238 \twoheadrightarrow 306$

$\downarrow$

$645 \twoheadleftarrow 387 \twoheadleftarrow 766 \twoheadleftarrow 725 \twoheadleftarrow 588 \twoheadleftarrow 571 \twoheadleftarrow 534 \twoheadleftarrow 525 \twoheadleftarrow 507 \twoheadleftarrow 854$

$\downarrow$

$910 \twoheadrightarrow 147 \twoheadrightarrow 437 \twoheadrightarrow 735 \twoheadrightarrow 864 \twoheadrightarrow 417 \twoheadrightarrow 734 \twoheadrightarrow 804 \twoheadrightarrow 365 \twoheadrightarrow 951$

the orbit of the binary arithmetic logarithm X. These are depicted below.

This diagram shows that the binary arithmetic logarithm of period 12 is almost the most complicated of all binary functions with period 12. This vertex X belongs to the component whose cycle has the largest period, and is located at almost the greatest distance from the cycle (which is equal to 3, while the treetops of the forest growing on this cycle are at the distance 4 from the cycle).

The entire theory described above remains empirical for now: neither its generalization to the cases $n > 12$, nor the proofs of the various conjectures stated above (for example, the conjecture about the great complexity of the binary arithmetic logarithm for greater values of n) are known.

Further, the theory of binary functions developed here (with values in $\mathbb{Z}_2$), that is, sequences of zeroes and ones, should be generalized to the case of functions with a larger number of values (for example, for functions with values in $\mathbb{Z}_p$ or in $\mathbb{Z}_n$).

This relates, for example, to the question of the complexity of the function $\ell(k)$ defined above with argument $k \in \mathbb{Z}_p \setminus 0$ and with values in $\mathbb{Z}_{p-1}$, which arise in the context of Fermat's little theorem. How disorderly (random) are the logarithms of a sequence of numbers (or, conversely, the elements of a geometric progression of residues)?

4. Complexity and Randomness of Tables of Galois Fields

Analogous questions about randomness also arise in the theory of Galois fields. The field $\mathbb{Z}_p$ is one example, but the general Galois field has p^k elements. (See Editor's note 4, page 81.) For example, even the field of p^2 elements will give us many seemingly random tables, but their complexity is not described, as of now, by any well-defined measure. It remains an empirically observed fact.

Here is an example of such "complexity" of a table of p^2 cells, filled with the numbers from 1 to p^2.

Let $A = \begin{pmatrix} a & b \\ c & d \end{pmatrix}$ be the matrix of order 2 whose elements are the remainders upon division by a prime p such that all the $p^2 - 1$ matrices A^k $(1 \leq k \leq p^2 - 1)$ are different, and such that $A^{p^2-1} = 1$.

This may happen: for any prime number p such a matrix exists. For example, for $p = 5$ the matrix

$$A = \begin{pmatrix} 0 & 2 \\ 1 & 2 \end{pmatrix}$$

will work.

For an arbitrary matrix $A = \begin{pmatrix} a & b \\ c & d \end{pmatrix}$, we know that $A^2 = \alpha A + \beta$ where $\alpha = \operatorname{Tr} A = a+d$ and $\beta = -\det A = bc - ad$. (For our example, $\alpha = 2$, $\beta = 2$). This yields an easy computation

$$A^3 = \alpha A^2 + \beta A = (\alpha^2 + \beta)A + \alpha\beta$$

and so on, so that $A^k = u_k A + v_k$ (where $0 \le u_k \le p$, $0 \le v_k \le p$).

Let us place the number k into the cell with coordinates (v_k, u_k) in a table of size $p \times p$. The resulting matrices A^k, together with the zero matrix (which corresponds to $k = \infty$, $u_\infty = v_\infty = 0$), form a Galois field of p^2 elements. The operations in this field are:

$$A^k \cdot A^\ell = A^{k+\ell}, \quad A^k + A^\ell = A^{k*\ell},$$

where $*$ is the operation of adding locations in the matrix:

$$v_{k*\ell} = v_k + v_\ell, \quad u_{k*\ell} = u_k + u_\ell.$$

The symbol A^∞ represents the zero matrix. In the case $p = 5$, the table constructed as described takes the form shown in Figure 4.

13	15	5	16	20
7	10	9	14	23
19	11	2	21	22
1	8	54	17	3
∞	24	18	6	12

Figure 4. The table for the Galois field of 25 elements.

Looking at these numbers, one can convince oneself that they fill the table "randomly". In trying to verify the "frequency" of various events that can be regarded as indicators of a "random" filling, the larger p is, the closer we get to the expected result.

Hypothetically this is asymptotically true for the appearance of the first N numbers of the table in any region (if it is defined in a way that is not too complicated): the expected values of these appearances is equal to ϑN, if the ratio of the area of the region to the area of the table is ϑ.

For example, let us take the first $N = 10$ of 25 values ($k = 1, \ldots, 10$). The first two of the five columns of the table comprise $\vartheta = 2/5$ of its area. Filling it randomly means that we expect $(2/5)N$ appearances. In the table, these values occur 4 times: $k = 1, 8, 7, 10$. Such a perfect agreement between the number of appearances and ϑN doesn't always happen, but the values become closer as the table grows, if the region under discussion is defined by an algorithm which is not too complicated.

Listeners may choose for themselves the criteria for "complexity", and verify their choices experimentally. A certain number of examples can be found in my article [**3**] and in the booklet [**9**].

The statement that the criteria of stochasticity (randomness) are fulfilled in the limit as $p \to \infty$ is well supported by experiments, but not always by proof, and I here relate a story about this, in the hopes of finding help (albeit experimental help) from my readers in their research into this question.

The following experiment relates to the theory of shuffling cards. We compare the table of a Galois field of p^2 elements to the following permutation of p^2 elements.

To each number k ($1 \leq k < p^2$) we assign the address $K = pu_k + v_k$ of that cell of the table containing the given number. For example, in the table of Figure 4 we obtain $K(1) = 5, K(2) = 12, \ldots, K(20) = 24$.

The conjecture is that this permutation is just as "random" as a good shuffling of a deck of $p^2 - 1$ cards (the quality of the shuffle increases as $p \to \infty$).

If desired, we can add one element, permuting all p^2 elements of the Galois field. To do this we must replace the symbol A^∞ with 0 (writing $k = p^2$ instead of ∞) and add the address $K = 0$ of the location of the zero matrix.

In the next lecture we will discuss how "random" the resulting permutation of $p^2 = 25$ elements is.

For Galois fields of p^n elements the table of the field fills a n-dimensional cube with edge p. The table is based on the expression of an $n \times n$ matrix A^k, whose elements are the remainders upon division by p, in the form of linear combinations of the n matrices $1, A, A^2, \ldots, A^{n-1}$:

$$A^k = u_k A^{n-1} + v_k A^{n-2} + \cdots + w_k A^{n-n}.$$

The "randomness" of the tables of such fields increases as $p \to \infty$ in the same way as described above for $n = 2$. Listeners can check this experimentally, although much of it has not yet been proven.

I will briefly discuss the connection between "complexity" and "randomness", which we've looked at in the present lecture, and complexity and randomness, which are defined completely differently in the theory of algorithms and probability theory.

From a statistical point of view, the general statistical laws (like the "Law of Large Numbers" about the approach of the frequency of successes in repeated trials to the probability of success in one trial) hold only for the *majority* of sequences of trials.

The majority of the sequences we study are also "complex" in our sense, because the larger part of the vertices of the corresponding graph are located on components with long cycles, and also on the highest branches of the trees attached to these cycles.

However, the conjecture states that the criteria for stochasticity are satisfied (and the longer the length n of the sequence considered, the greater the accuracy of the stochasticity), not just for typical sequences (i.e., for the majority of sequences), but even for atypical sequences that are "complicated" in our sense.

Many statistically "typical" objects can turn out to be uncomplicated in our sense, but these, presumably, make up the minority (just as do statistically atypical sequences which are "complicated" in our sense, if they exist).

We can also assert that finite models of algorithmically non-computable sequences turn out, as a rule, to be more complicated in our sense, than the analogous finite models of algorithmically computable sequences.

However, this hypothesis is not only unproven, but not even formulated as of yet in a precise way, so that one might attempt a proof. (I expect both a formulation and a proof from my listeners: this is why I am telling them this story.)

Many mathematicians think that one can understand a theorem simply by generalizing it, so that the regularities found can be spread to a greater circle of phenomena.

For this reason I have not stopped at the description above of the theory of complexity of sequences of binary digits. I have also done analogous experiments for sequences consisting of other objects. For example, I have experimented with ternary sequences, consisting of remainders upon division by 3 (that is, replacing the remainders $\{0,1\}$ upon division by 2 with the remainders $\{0,1,2\}$ upon division by 3).

Here are the simplest results of these experiments (relating to sequences of $n \leq 7$ symbols, forming a set of 3^n elements).

The operator $A = \delta - 1\colon \mathbb{Z}_3^n \to \mathbb{Z}_3^n$ acts according to the previous formula

$$(A_x)_k = x_{k+1} - x_k \text{ (where } x_{n+1} = x_1).$$

A direct computation of the graphs of these operators leads us to the following table of results:

n	b	components of the graph	relation
2	1	$(O_1 * T_3)$	$A^2 = A$
3	1	$(O_1 * T_{27})$	$A^3 = 0$
4	6	$3(O_8 * T_3) + 3(O_1 * T_3)$	$A^9 = A$
5	2	$(O_{80} * T_3) + (O_1 * T_3)$	$A^{81} = A$
6	11	$8(O_3 * T_{27}) + 3(O_1 * T_{27})$	$A^6 = A$
7	3	$2(O_{364} * T_3) + (O_1 * T_3)$	$A^{365} = A$

In the ternary trees (T_3 and T_{27}) each vertex except the highest has three preimages:

I invite the reader to carry over the investigation of the previous theorems on binary sequences to the case of remainders upon division

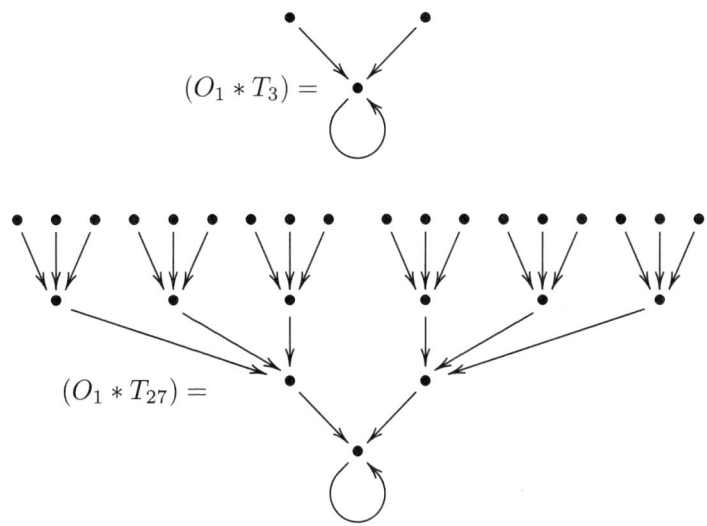

by p (for example, by studying the "polynomial tree" $(O_1 * T_{3^k})$ in each component of the previous table).

The period $T = 3$ of the longest cycle of the graph of the operation $A \colon \mathbb{Z}_3^6 \to \mathbb{Z}_3^6$ (in the case $n = 6$) is not divisible by n, so the generalization of the corresponding property of binary sequences is not trivial.

The case $n = 7$ of our table is clearly related to the astronomy of a year of $364 = (3^6 - 1)/2$ days and $364/28 = 13$ months.

My Dubna lecture of 2005 gave rise to a continuing investigation into it by O. Kaprenkov and A. Garber. The latter, in particular, discovered that if $n = p - 1$ is not divisible by 8, then the "logarithm" is complex (for example, for $p = 4k + 3$), but for $p = 73$ the "logarithm" belongs to the cycle. O. Karpenkov found that the fraction T/n is not always one less than a power of 2. For example, for $n = 23$, $T = 2047$, we have $T/n = 89$. I hope that all these results will be published soon. (See Editor's note 5, page 81.) See also [7].

Editor's notes

1. One classical measure of randomness is called *Kolmogorov's complexity*. The complexity of a sequence of zeroes and ones (or,

actually, of any other object) can be measured by the minimal length (that is, the minimal number of letters and digits) of its description in some language, say, in English. For example, the sequence

$$\underbrace{001\ 001\ \cdots\ 001}_{100}$$

can be described by a sentence "001 repeated 100 times" which has 19 letters (22, if you count letters and spaces). This is a relatively small number, so the sequence has a low complexity (low randomness).

It is interesting that this transparent definition is not as simple as it seems to be. It may lead to unexpected paradoxes. For example, does the sentence "The smallest positive integer which cannot be described by a sentence of less than 300 letters" describe any number? Seemingly, yes: all sequences of less than 300 letters and spaces form a finite set. Some of them describe positive integers (for example, "five hundred thirty one" or "the population of Los Angeles"), while others (the vast majority) do not. Thus, there are finitely many positive integers which are described in our way, and there is a smallest positive integer which does not have this description. This is our number. But it *has* a description (given above) by a sentence of 93 letters, digits and spaces (if my count is correct). We arrive at a paradoxical contradiction.

I do not know who was the first to observe this amusing paradox; I first heard of it in the mid-1970's from N. Konstantinov.

The reader should not worry: Arnold's measure of complexity/ randomness is completely different and (hopefully) free of paradoxes.

2. A composite integer a is called a *base 2 pseudoprime*, if $2^a \equiv a \bmod 2$. The number 341 is the first example of a base 2 pseudoprime. There are also base 3, base 5, etc. pseudoprimes. A number a is called a *Carmichael number*, if it is a pseudoprime with respect to any base; the smallest Carmichael number is 561. The reader can find extensive information about pseudoprimes, Carmichael numbers and similar further notions on the web. I can also recommend the book of K. Rosen [**18***].

3. Differences provide a discrete version of derivatives. In this way, the approach to complexity of a function based on studying the behavior of its differences may be regarded as a discretization of a similar approach in analysis where a function is considered less complex if there a simple relation between its derivatives; that is, if it satisfies a simple differential equation. For example, a polynomial $y = p(x)$ of degree n is not complex, since it satisfies a simple differential equation $y^{(n)} = 0$. The functions $y = e^x$ and $y = \sin x$ satisfy differential equations $y' = y$ and $y'' + y = 0$, so they are also not complex. The function $y = \sqrt{x}$ is more complex: the simplest differential equation it satisfies, $2yy' = 1$, is not linear. It is still worse for a logarithmic function, which appears to be very complex; this is what Arnold demonstrates in the section below.

4. A *field* is a set with two binary operations called addition and multiplication which satisfy the ordinary properties of commutativity, associativity, distributivity, and the existence of additive and multiplicative inverses. Galois fields are fields with finitely many elements. A well known fact, proved in many textbooks in algebra, is that the number of elements of a Galois field is a power of a prime, p^k, and for any p and k there exists a Galois field with p^k elements which is unique up to isomorphism. A Galois field with a prime number p of elements may be constructed as the field $\mathbb{Z}_p$ of residues modulo p. There are some other explicit constructions of Galois field; for example, if $p \equiv 3 \bmod 4$, then the "complex numbers" $a + b \cdot i$ with $a, b \in \mathbb{Z}_p$ and with the usual addition and multiplication (in particular, $i^2 = -1$) comprise a field with p^2 elements. However, an explicit (tabular) description of a Galois field (even with p^2 elements) has a high level of complexity, and this is the subject of Section 2.4.

5. At least, some of these results have been published. See [13*] and [14*].

Lecture 3

Random Permutations and Young Diagrams of Their Cycles

"Everyone considers himself an expert on what he knows least about."

Oscar Wilde

"A published manuscript is comparable to a whore."

Jonathan Swift, according to V. Nabokov

Every permutation of n elements can be partitioned into cycles. For example, for the following permutation which takes the digits x into the digits y:

x	0	1	2	3	4	5	6	7	8	9
y	8	4	2	5	3	1	6	9	0	7

there are 5 cycles, namely

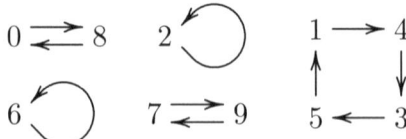

This we obtain a partition of the number n into the lengths of cycles: $10 = 4 + 2 + 2 + 1 + 1$. The partition of a natural number into natural addends is usually depicted as its "Young diagram", which in this case has the form

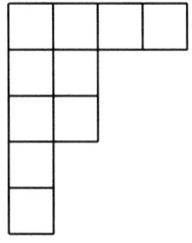

For the partition $n = x_1 + \cdots + x_y$, $x_1 \geq x_2 \geq \cdots \geq x_y$ the first row has x_1 cells, the second row has x_2 in the second cell, and so on, until we reach the shortest of the y rows.

The numbers $x = x_1$ and y are called the *length* and the *height* of the diagram. The Young diagram of the partition of ten digits into cycles for our permutation thus has length $x = 4$ and height $y = 5$. In our example, it occupies an area of n cells and a fraction $\lambda = n/(xy)$ of the area of its circumscribed rectangle (with sides x and y). In our example, $\lambda = 10/20 = 1/2$. We will call the number λ the *fullness* of the diagram.

In this lecture we will investigate Young diagrams for several special permutations.

There are a total of $n!$ permutations of n elements, and it is interesting to ask which Young diagrams (of the partition into cycles) occur more frequently among these $n!$ "random" permutations, and which more rarely, in particular, what is the average length, height, or fullness of the Young diagram, and what is usually larger, the length or the height.

For small n these questions can be answered by drawing all the Young diagram of area n for all $n!$ permutations, but the number p_n of these grows large as n increases:

n	1	2	3	4	5	6	7	8	9	10
p_n	1	2	3	5	7	11	15	22	30	42

(See Editor's note 1, page 106.)

For large values of n it is simpler to go the route of heuristic experiments: start with an artificial "random" permutation of n elements (say, $n = 100$), and construct its Young diagram. (I propose that the Young diagrams of 100! permutations of 100 elements in most cases resemble one another. This is of course not obvious, nor is the fact that one artificially defined permutation is typical of them all). To test whether it is typical, we must repeat the experiment a number of times, and compare the results.

Aside from this, we compare "random" permutations with the permutations of points in dynamic systems with a finite phase space, which are provided us from algebra and number theory. (These systems also model finite cycles of periodic points in dynamical systems in discrete time).

1. Statistics of Young Diagrams of Permutations of Small Numbers of Objects

For small n it is not hard to count all the partitions of n. For example, for $n = 3$ there are only 3 of them, with Young diagrams

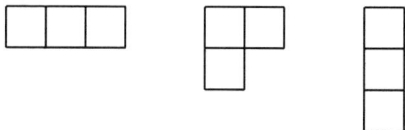

These represent the partitions $3 = 2+1 = 1+1+1$.

For brevity of notation, I will write the partition $n = x_1 + \cdots + x_y$ in the form of a "monomial":

$$a^\alpha b^\beta \cdots, \quad a > b > \cdots,$$

where the largest addend $x_1 = a$ is repeated α times in the partition, the second largest addend is repeated β times, and so on.

The three partitions of $n = 3$ that we've listed correspond to the monomials $D = 3, 2 \cdot 1$, and 1^3 (where we have not written an exponent 1).

The group $S(3)$ of six permutations of three elements can be seen as the group of symmetries of an equilateral triangle (by thinking of these as permuting the three vertices).

This group consists of the identity transformation, two rotations (of $120°$ and $240°$) of order 3 and three line symmetries of order 2 (reflections in the three altitudes of the triangle).

The Young diagrams of these transformations are, respectively

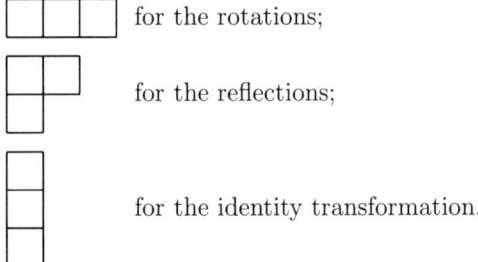

for the rotations;

for the reflections;

for the identity transformation.

We will write these conclusions in a table.

The case $n = 3$. *The classes of conjugate elements of the symmetric group $S(3)$ (the group of symmetries of an equilateral triangle).*

D	3	$2 \cdot 1$	1^3
x	3	2	1
y	1	2	3
N	2	3	1

The row marked N shows the number of permutations with a given Young diagram D. All these permutations, for a given Young diagram, are "alike" (or "similar", or, as the algebraists say, "conjugate" in the group $S(n)$ of permutations of n elements) and differ only in the names of the elements being permuted: they are "indistinguishable from a relativistic point of view". (See Editor's note 2, page 107.)

The total number of permutations is the sum over all diagrams with a given area n:

$$n! = \sum_D N(D),$$

since each permutation can be partitioned into cycles in just one way.

Counting the number $N(D)$ of representations by permutations for a given Young diagram D is an elementary problem in combinatorics; all we need is to count the number of ways to re-order the symbols $(1, 2, \ldots, n)$ in the cells of a Young diagram which yield different permutations.

For a diagram with one cycle of length n (where $D = n$) the number of different permutations is $N(n) = (n-1)!$; for example, $N(4) = 6$.

Indeed, a cycle must include the element 1, after which must follow one of the $(n-1)$ remaining elements, then one of the remaining $(n-2)$ elements, and so on. Thus the number of different cyclic permutations of a set of n elements is equal to $(n-1)(n-2)\cdots = (n-1)!$.

Listeners can easily prove themselves the results about permutations of $n \leq 7$ elements, analogous to those given above for a table where $n = 3$.

Theorem 1. *Partitions of the groups $S(n)$ of permutations of n elements into classes of conjugate elements (with Young diagrams of length x and height y for $N(D)$ permutations which are conjugate to one another) are provided, for $n \leq 7$, in the following tables.*

The case $n = 4$. *The classes of conjugate elements of the symmetric group $S(4)$ (the group of symmetries of a tetrahedron).*

D	4	$3 \cdot 1$	2^2	$2 \cdot 1^2$	1^4
x	4	3	2	2	1
y	1	2	2	3	4
N	6	8	3	6	1

The column $N(2^2) = 3$ corresponds to the three axes of symmetry of a tetrahedron, connecting the midpoints of opposite (skew) edges. These three symmetries of a tetrahedron, together with the identity transformation, form a wonderful commutative subgroup $\mathbb{Z}_2 \times \mathbb{Z}_2 \subset S(4)$, thanks to which algebraic equations of the fourth degree are solvable in radicals.

The case $n = 5$. *The classes of conjugate elements of the group $S(5)$.*

D	5	$4 \cdot 1$	$3 \cdot 2$	$3 \cdot 1^2$	$2^2 \cdot 1$	$2 \cdot 1^3$	1^5
x	5	4	3	3	2	2	1
y	1	2	2	3	3	4	5
N	24	30	20	20	15	10	1

This table demonstrates the entire geometry of the group of symmetries of a dodecahedron, with its 12 faces, 20 vertices, and 30 edges.

Each face gives rise to two cyclic rotations of order five. Each vertex yields rotations of order 3 which keep the vertex fixed. Each edge gives rise to a rotation of order 2 which fixes a pair of opposite edges.

To interpret the symmetries of a dodecahedron as permutations of 5 elements, Kepler inscribed five cubes in a dodecahedron, whose edges are diagonals of the faces of the dodecahedron. Here is how they are constructed (Figure 1).

Having chosen a diagonal d of one of the faces, we can draw two diagonals d', d'' of two neighboring faces, through one of its endpoints O, which are obtained from the diagonal d by rotations of the dodecahedron which fix the chosen endpoint. The three diagonals obtained in this way turn out to be orthogonal to each other.

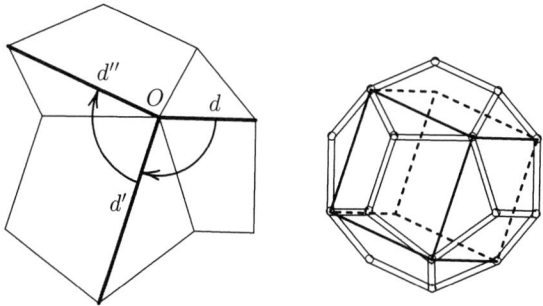

Figure 1. Kepler's cube inscribed into the dodecahedron.

We then repeat this construction at the free endpoints of the diagonals we have already obtained, to construct new diagonals of the faces, until we have 12 diagonals which form a cube (with one edge on each of the 12 faces of the dodecahedron).

Depending on which of the five diagonals of the original face we choose to start this construction, we will obtain Kepler's five cubes, each inscribed in the dodecahedron, which are also permuted by the group of symmetries of the dodecahedron.

The case $n = 6$. *The classes of conjugate elements of the group $S(6)$.*

D	6	5·1	4·2	4·1²	3²	3·2·1	3·1³	2³	2²·1²	2·1⁴	1⁶
x	6	5	4	4	3	3	3	2	2	2	1
y	1	2	2	3	2	3	4	3	4	5	6
N	120	144	90	90	40	120	40	15	45	15	1

The case n=7. *The classes of conjugate elements of the symmetric group $S(7)$.*

D	7	6 · 1	5 · 2	5 · 1²	4 · 3	4 · 2 · 1	4 · 1³
N	720	840	504	504	420	630	210

D	3² · 1	3 · 2²	3·2·1²	3 · 1⁴	2³ · 1	2² · 1³	2 · 1⁵	1⁷
N	280	210	420	70	105	105	21	1

Let us prove, for example, that $N(3^2 \cdot 1) = 280$. A cycle of length 1 may consist of any of the 7 elements, and the choice of this element is determined uniquely by a permutation with the diagram $D = 3^2 \cdot 1$ uniquely. Then we need to divide the other 6 elements into two cycles of 3 elements each. One triple can be chosen in $\binom{6}{3} = 20$ ways, but there are only 10 different ways to divide the the set of 6 elements into triples, since we don't know which of the two triples is the first.

Once the division into cycles is fixed, it remains to choose the cyclic order in each of the triples (each independent of the other). The number of cyclic orders of $m = 3$ elements is $(m-1)! = 2$. Thus there are $2 \cdot 2 = 4$ cyclic orders in both triples, and altogether there are $4 \cdot 10 = 40$ permutations of our type.

Taking into account the arbitrariness of the choice of the seventh element, which is fixed by the permutation, we find that the number of all permutations with diagram $D = 3^2 \cdot 1$ is equal to the product $40 \cdot 7 = 280$.

The other entries in the table can be computed in a similar way.

The tables we have given allow us to compute (for $n \leq 7$) the average $\widehat{f}$, over all $n!$ permutations of n elements, for any function f of the Young diagram D with area n:

$$\widehat{f}(n) = \sum_{D} (N(D)f(D))/n!$$

Theorem 2. *The average values of the characteristics of a diagram (its length x, its height y, its fullness $\lambda = n/(xy)$, its asymmetry $\mu = y/x$) for $n \leq 7$ have the values shown in the table on the next page.*

Although we have obtained these statistics only for rather small values of n, they hint at a whole range of conjectures.

For example, we can assert that the average length of a diagram grows approximately linearly as its area n grows: $\widehat{x} \approx c_1 n$ (the table suggests that the value of the coefficient c_1 is close to $2/3$).

Despite this, the average height of a diagram grows rather slowly as its area grows, presumably even

$$\widehat{y} \sim c_2 \ln n$$

n	$\hat{x}$	$\hat{y}$	$\hat{\lambda}$	$\hat{\mu}$
2	$3/2 = 1.50$	$3/2 = 1.50$	1.00	$5/4 = 1.25$
3	$2\frac{1}{6} \approx 2.17$	$1\frac{5}{6} \approx 1.83$	$7/8 \approx 0.88$	$10/9 \approx 1.11$
4	$2\frac{19}{24} \approx 2.79$	$2\frac{1}{12} \approx 2.08$	$\frac{29}{56} \approx 0.80$	$\frac{137}{144} \approx 0.95$
5	$3\frac{17}{40} \approx 3.42$	$2\frac{17}{60} \approx 2.28$	$\frac{325}{432} \approx 0.75$	$\frac{184}{225} \approx 0.82$
6	$4\frac{31}{720} \approx 4.04$	$2\frac{408}{726} \approx 2.57$	≈ 0.72	≈ 0.76
7	≈ 4.68	≈ 2.71	≈ 0.69	≈ 0.70

(the height is approximately doubled when the area is squared).

The average coefficient of fullness of a diagram slowly decreases as the area n increases. The table suggests that it behaves like

$$\hat{\lambda} \sim \frac{c_3}{\ln n},$$

(with the fullness approximately halved when the area n is squared).

The size of the average asymmetry μ decreases in the sense that as the area n increases, typical diagrams get flatter and become lower and longer.

However, the quadratic mean of the value of the logarithm of asymmetry μ is $\sigma \approx 0.30$ for $n = 2$ and grows to $\sigma \approx 0.42$ for $n = 7$. This growth in the size of σ shows that a rather significant asymmetry remains in a significant fraction of the diagrams even for large areas n; that is, that the diagrams do not all get more symmetric as the area increases. On the contrary, they assume a wide variety of forms, including long diagrams with $\mu < 1$ and tall diagrams with $\mu > 1$. (The quadratic mean of the value of σ does not distinguish between these two types of asymmetries.)

2. Experimentation with Random Permutations of Larger Numbers of Elements

Computation by computer could be continued (the computation of the average characteristics of Young diagrams of permutations with values of n larger than those discussed in Section 3.1). However, the summations required for all $n!$ permutations of n elements are not realistic (even using a computer) for values of n such as 100: we are not able to add 100! numbers.

So I invented another approach to this problem (which is more like natural science than mathematics): instead of summing over all 100! permutations, I used one permutation of 100 elements, which I chose in a perfectly random fashion. I regarded the characteristics of its Young diagram showing the partition of the number $n = 100$ into the lengths of cycles as typical characteristics of a diagram with area n of a "random" permutation.

My way of constructing a "random" permutation of n elements was as follows. (I describe it below using $n = 100$, to simplify the notation.)

We start with some sequence of random digits. I will describe below several sources of such sequences: we used the telephone numbers of members of the National Academy of Sciences of the USA in one case, and the license plate numbers of passing cars on Vavilov Street past the Mathematical Institute of the Russian Academy of Sciences in another, and the results were similar.

In a random permutation of digits $\alpha\,\beta\,\gamma\,\delta\,\cdots$ we regard $(\alpha\beta)$ as a two-digit number. Then we begin our permutations (that is, our ordering) of all 100 two-digit numbers (00, 01, ,..., 99) with the number $(\alpha\beta)$. If the number $(\gamma\delta)$ is different from $(\alpha\beta)$, then we us it as the second member of the sequence of two-digit numbers we are constructing. If it is not different, we skip it and go to the next number, $(\varepsilon\zeta)$. And so on: if some of the first elements of the permuted sequence were already chosen, then instead of using it for the next element, we take the first of the following pairs of elements of the original sequence of random digits which differs from all those already taken.

In the end, this algorithm will construct for us a random permutation of all 100 two-digit numbers. But as we get to the end of the process, it works slower and slower, because before we encounter a new two-digit number (which has not be chosen yet), we must move a long way through candidates which have already been chosen, looking to see if they are new, one after another.

To be specific, if we were to permute n elements, then the number of candidates we would have to examine would be of the order $n \ln n$, as explained below. For $n = 100$ we find $\ln 100 \approx 4.6$; that is, we would have to work with a table of approximately 500 random two-digit numbers.

The list of academicians is sufficiently long for this, and I have chosen, as "random" two-digit numbers, pairs of digits which are the fourth and fifth in the seven-digit telephone numbers of the academicians (in alphabetical order, as they appear in the phone directory).

The following protocol for such an experiment consists of a row of the chosen two-digit numbers, with rejected candidates in parentheses. In this experiment of 100 trials, 64 two digit numbers were chosen, and the frequency of repeated candidates slows the process down significantly towards the end.

47	99	07	32	02	91	52	66	21	81
27	82	70	43	17	65	76	28	63	08
94	11	01	95	(52)	(76)	87	(65)	29	16
20	80	10	25	37	(65)	(32)	35	(21)	74
05	36	48	(24)	73	(48)	90	18	75	12
(02)	15	41	72	38	61	(73)	(73)	(63)	(11)
24	83	56	(32)	(74)	06	84	(56)	(81)	67
14	03	(83)	(56)	96	(48)	(27)	(37)	97	(08)
(37)	89	(02)	(97)	(38)	(52)	44	19	(24)	(28)
(12)	(01)	13	69	(20)	(17)	(84)	88	53	(61)

To counter this slowness at the end of the process, I invented a few more enhancements. First of all, instead of n elements, we can choose only $n/2$, then choose $n/2$ more elements from the remaining $n/2$ candidates, using the same method. For example, we can somehow bijectively map the remaining set of $n/2$ permuted elements onto the set of elements already ordered, and then order them using the $n/2$ chosen elements which have already been ordered.

For another method, we can choose the second half of the set of random two-digit numbers from the sequence $(\beta\gamma)(\delta\epsilon)\cdots$, rather than from the original sequence $(\alpha\beta)(\gamma\delta)\cdots$.

A third method consists in using, for the ordering of $m = 100/k$ elements, the remainders upon division of the original two-digit numbers $(\alpha\beta)(\gamma\delta)\cdots$ by the number m which is a factor of 100.

Each of these methods allows us to quickly put all $n = 100$ two-digit numbers in a "random" order, which gives us the necessary "random" permutation of a set of n elements.

For $n = 16$, using a random sequence of remainders upon division by 16 in this way, I obtained the following permutation of the elements $\{0, 1, \ldots, 15\}$:

$$0\ 4\ 3\ 12\ 9\ 8\ 7\ 14\ 5\ 1\ 2\ 11\ 6\ 15\ 10\ 13.$$

The cycles corresponding to this permutation are easily computed:

$0 \to (0)$, of length 1;

$1 \to 4 \to 9 \to (1)$, of length 3;

$2 \to 3 \to 12 \to 6 \to 7 \to 14 \to 10 \to (2)$, of length 7;

$5 \to 8 \to (5)$, of length 2;

$11 \to (11)$, of length 1;

$13 \to 15 \to (13)$, of length 2.

Thus the Young diagram of our "random" permutation of 16 elements is $D = 7 \cdot 3 \cdot 2^2 \cdot 1^2$.

The parameters of this diagram have the values

$$x = 7, \quad y = 6, \quad \lambda = \frac{16}{42} \approx 0.38, \quad \mu = \frac{6}{7} \approx 0.86.$$

They fit comfortably into the conjectures about the behavior of the average characteristics which we predicted in the end of Section 3.1.

Remark. The expression $n \ln n$ for for the number of attempts we make to randomly arrange n elements has the following heuristic explanation.

The last of the n chosen elements occur as we continue our random sequence with probability $1/n$, since we can expect its appearance on the average after n attempts.

For the choice of the penultimate element, we will have success if we encounter any of the two elements not yet chosen from the n possible elements. The expected number of trials must then be $1/n$.

The choice of the previous elements requires, on the average, $n/3$ attempts, and so on. The expected average total number of attempts is:

$$n \left(\sum_{k=1}^{n} \frac{1}{k} \right) \sim n \ln x |_1^n \sim n \ln n$$

(We have used the fact that $\displaystyle\int \frac{dx}{x} = \ln x$).

Similar heuristic arguments show why the expected number of cycles in a permutation of n elements is of order $c_2 \ln n$.

It is natural to think that in a random sequence of independent choices of n elements the first repetition will come after cn attempts, where c is some constant. This gives a expected length cn for the first cycle.

From that point the choice is made from $n_1 = n - cn = (1 - c)n$ candidates.

Thus after forming y cycles, the number of remaining candidates will be $n_y = (1 - c)^y n$.

But this whole procedure ends with n_y of order 1. This relationship gives us the following expression for the number y of cycles:

$$y \ln(1 - c) + \ln n \approx 0, \quad y \approx \frac{\ln n}{\ln(1 - c)^{-1}} \sim c_2 \ln n.$$

Having performed rather a large number of similar experiments, I obtained Young diagrams for "random" permutations of n elements with characteristics shown in the table on the next page.

All these characteristics of empirically "random" permutations agree well with the conjectures of Section 3.1. But of course further experimental verification (especially with large values of n) would be

n	D	x	y	λ	μ
16	$7 \cdot 3 \cdot 2^2 \cdot 1^2$	7	6	0.38	0.86
25	$9 \cdot 7 \cdot 5 \cdot 3 \cdot 1$	9	5	0.56	0.56
64	$35 \cdot 15 \cdot 7 \cdot 3 \cdot 2 \cdot 1$	35	7	0.26	0.20
100	$42 \cdot 36 \cdot 18 \cdot 2 \cdot 1$	42	6	0.40	0.14
100	$90 \cdot 4 \cdot 3 \cdot 2 \cdot 1$	90	5	0.22	0.06
100	$88 \cdot 9 \cdot 1$	88	5	0.23	0.06
169	$147 \cdot 13 \cdot 8 \cdot 1$	147	4	0.29	0.03

desirable, and would not require any previous knowledge, so would be accessible to secondary school students.

I performed one more independent verification choosing as a source of random permutations the table of the Galois field of p^2 elements.

3. Random Permutations of p^2 Elements Generated by Galois Fields

The construction of these permutations was explained in Section 2.4. Recall that a Galois field of p^2 elements consists of matrices

$$0 \text{ and } A^k, \ 1 \leq k < p^2, \text{ where } A^k = u_k A + v_k,$$

where $A = \begin{pmatrix} a & b \\ c & d \end{pmatrix}$ is a matrix whose elements are the remainders upon division of the prime number p and such that

$$A^{p^2-1} = 1, \ A^{k<p^2-1} \neq 1.$$

A permutation of a set of p^2 elements $1 \leq k \leq p^2$ is defined by this field as follows: for $k < p^2$, we define the "address" $K = u_k p + v_k$ of the matrix $A^k = u_k A + v_k$ in the table of the field, and for $k = p^2$ we set $K = 0$ (so that $u_{p^2} = v_{p^2} = 0$).

Note that the addresses of the elements of the table run through the set $0 \leq K \leq (p-1)p + (p-1) = p^1 - 1$ of p^2 points (so that we can interpret both k and K as remainders upon division by p^2).

The transformation $k \mapsto K$ is a permutation of this set of p^2 elements, and we can apply the preceding theory to this transformation: we can partition it into cycles, construct the Young diagram, find its characteristics (its length x, its height y, its fullness λ, and its asymmetry μ).

After performing these computations for $p \leq 13$, I got surprising results, resembling the Young diagrams of random permutations. These results are given in the next table.

Theorem 3. *Young diagrams of permutations from tables of Galois fields of p^2 elements are as follows:*

p	n	D	x	y	λ	μ
3	9	$6 \cdot 2 \cdot 1$	6	3	0.50	0.50
5	25	$14 \cdot 5 \cdot 4 \cdot 1^2$	14	5	0.36	0.36
7	49	$16 \cdot 11 \cdot 7 \cdot 6 \cdot 4 \cdot 3 \cdot 1^2$	16	8	0.38	0.50
11	121	$65 \cdot 39 \cdot 5 \cdot 3^3 \cdot 2 \cdot 1$	65	8	0.25	0.12
13	169	$98 \cdot 55 \cdot 12 \cdot 2 \cdot 1^2$	98	6	0.29	0.06

A comparison of this table with previous tables can be considered as confirmation of the conjecture that tables of Galois fields for large values of p, have the statistical properties of tables of random numbers. (For example, the permutations these tables give of elements of the field have the properties of typical random permutations.)

Besides the empirical data we have given, there are no theoretical foundations, and this is one of the reasons I've included a description of these experiments in the present lecture.

On the other hand, the tables given here give a new argument in favor of the conjectures of Section 3.1 (if we agree to consider permutations given by a table of a Galois field as random).

4. Statistics of Cycles of Fibonacci Automorphisms

For another algebraic example of permutations of finite sets we will consider periodic points of special dynamical systems on a two dimensional torus, the so-called hyperbolic automorphisms.

An automorphism of a torus $T^2 = \mathbb{R}^2/\mathbb{Z}^2$ is given by a matrix $A \in SL(2,\mathbb{Z})$ with determinant 1. The linear transformation $A : \mathbb{R}^2 \to \mathbb{R}^2$ takes the lattice of integer vectors $\mathbb{Z}^2$ into itself, and therefore the diffeomorphism takes the torus into itself as well.

Periodic points of an automorphism (which we will call A, as before) are solutions of the equation

$$A^k x = x \ (\in \mathbb{T}^2).$$

We will suppose that they are isolated; that is, that $\det(A^k - 1) \neq 0$. In this case the periodic points belong to the finite tori

$$M \cong \mathbb{Z}_n^2,$$

consisting of points with rational coordinates

$$z = (z_1, z_2), \ \ Z_j = \frac{u_j}{n}, \ \ u_j \in \mathbb{Z}_n = \mathbb{Z}/(n\mathbb{Z}).$$

The number of such points (with a given denominator n) is finite:

$$|\mathbb{Z}_n^2| = n^2.$$

The mapping A permutes the points of a finite set M, and this permutation can be partitioned into cycles.

We now plan to study the Young diagrams of these partitions.

Consider the special *Fibonacci automorphisms* (which are also connected with the golden section), corresponding to the matrix $A = \begin{pmatrix} 2 & 1 \\ 1 & 1 \end{pmatrix}$:

(1) $$A \begin{pmatrix} z_1 \\ z_2 \end{pmatrix} = \begin{pmatrix} 2z_1 + z_2 \\ z_1 + z_2 \end{pmatrix}.$$

To simplify the notation, we can multiply the (fractional) coordinates by the common denominator n; that is, we can consider the components z_1 and z_2 as remainders upon division by n:

$$\begin{pmatrix} z_1 \\ z_2 \end{pmatrix} \in M = \mathbb{Z}_2.$$

The Young diagram of an automorphism $A : M \to M$ of a finite torus M has area n^2, and we propose to study its shape for the Fibonacci automorphism for various M.

Remark. We call the automorphism A a Fibonacci automorphism because it acts on the basis vectors in the following way:

$$\begin{pmatrix}0\\1\end{pmatrix} \mapsto \begin{pmatrix}1\\1\end{pmatrix} \mapsto \begin{pmatrix}3\\2\end{pmatrix} \mapsto \begin{pmatrix}8\\5\end{pmatrix} \mapsto \begin{pmatrix}21\\13\end{pmatrix} \mapsto \begin{pmatrix}55\\34\end{pmatrix} \mapsto \cdots .$$

The components of these vectors form the Fibonacci sequence 1, 1, 2, 3, 5, 8, 13, ..., and their ratio approaches the golden ratio $\dfrac{\sqrt{5}-1}{2} \approx$ 0.6 (since eigenvalues of the matrix A are equal to $\dfrac{3 \pm \sqrt{5}}{2}$).

We will also say that the diffeomorphism $A : T^2 \to T^2$ makes catsup out of a cat, since the image of a cat C drawn on T^2 after several iterations of the transformation A, which preserves areas, is "spread onto the torus" very evenly (Figure 2).[1]

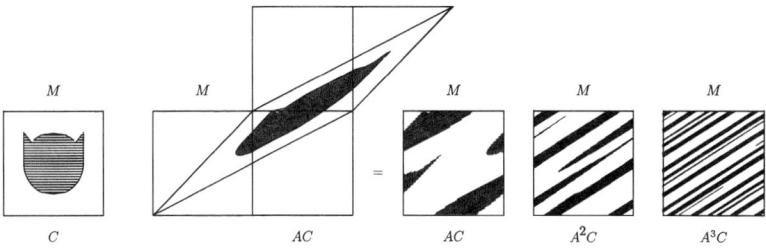

Figure 2. Making catsup from a cat

Direct computation of the iterations of the transformation A yields

Theorem 4. *The Young diagram D of the action of the Fibonacci automorphism (1) on the finite torus $\mathbb{Z}_n^2$ for $n \le 20$ has the characteristic shown in the table in next page.*

Remark. This theorem suggests a range of conjectures, for example, for prime values of $n = p > 5$ the diagrams have a standard simple form whose symbol is $D = x^z 1$, $z = y - 1$. The multiplicative character nature of D also becomes apparent: $D(pq)$ is related to $D(p)$ and $D(q)$.

---

[1]In the literature, the transformation A is commonly called *Arnold's cat map* – Editor's note.

n	D	x	y	$\lambda \approx$	$\mu \approx$
2	$3 \cdot 1$	3	2	0.67	0.67
3	$4^2 \cdot 1$	4	3	0.75	0.75
4	$3^5 \cdot 1$	3	6	0.89	2.00
5	$10^2 \cdot 2^3 \cdot 1$	10	5	0.50	0.50
6	$12^2 \cdot 4^2 \cdot 3 \cdot 1$	12	6	0.50	0.50
7	$8^6 \cdot 1$	8	7	0.88	0.88
8	$6^8 \cdot 3^5 \cdot 1$	6	14	0.76	2.33
9	$12^6 \cdot 4^2 \cdot 1$	12	9	0.75	0.75
10	$30^2 \cdot 10^2 \cdot 6^2 \cdot 3 \cdot 2^2 \cdot 1$	30	10	0.33	0.33
11	$5^{24} \cdot 1$	5	25	0.97	5.00
12	$12^{10} \cdot 4^2 \cdot 3^5 \cdot 1$	12	18	0.67	1.50
13	$14^{12} \cdot 1$	14	13	0.93	0.93
14	$24^6 \cdot 8^6 \cdot 3 \cdot 1$	24	14	0.58	0.58
15	$20^8 \cdot 10^2 \cdot 4^{10} \cdot 2^2 \cdot 1$	20	23	0.49	1.15
16	$12^{16} \cdot 6^8 \cdot 3^5 \cdot 1$	12	30	0.71	2.50
17	$18^{16} \cdot 1$	18	17	0.94	0.94
18	$12^{26} \cdot 6^8 \cdot 3^5 \cdot 1$	12	30	0.90	2.50
19	$9^{40} \cdot 1$	9	41	0.98	4.80
20	$30^{10} \cdot 10^2 \cdot 6^{10} \cdot 3^5 \cdot 2^2 \cdot 1$	30	30	0.44	1.00
41	$20^{84} \cdot 1$	20	20	0.99	4.25
97	$98^{96} \cdot 1$	98	98	0.99	0.99

I leave a precise formulation of this range of conjectures to the readers (hoping that they will, in time, prove some of them).

Several special diagram for several special automorphisms have been computed earlier by Percival, Vivaldi, Dyson and others. But their special examples make up only a small part of all possible cases, and I would not dare to use them to make any conclusions about typical diagrams of the lengths of cycles (even in the case of the Fibonacci automorphism for a typical value of n).

Together with the finite torus $M = \mathbb{Z}_p^2$ we can also look at a finite projective line

$$P = P^1(\mathbb{Z}_p) = (\mathbb{Z}_p^2 \setminus 0)/(\mathbb{Z}_p \setminus 0),$$

which consists of $p + 1$ points.

The linear operator $A : M \to M$ acts on P as a specific (projective) permutation of these points:

$$A_p \in GP(\mathbb{Z}_p) \subset S(p + 1).$$

The partition of this projective line into cycles of a projective permutation A_p defines its Young diagram (whose area is $p + 1$).

Theorem 5. *The Young diagrams of the cycles of the projective permutations A_p, generated by the Fibonacci automorphisms $A(z_1, z_2) = (2z_1 + z_2, \ z_1 + z_2) \bmod p$, have for prime $p < 20$ the characteristics shown in the table below:*

p	D	x	y	$\lambda \approx$	$\mu \approx$
2	3	3	1	1.00	0.33
3	2^2	2	2	1.00	1.00
5	$5 \cdot 1$	5	2	0.60	0.40
7	4^2	4	2	1.00	0.50
11	$5^2 \cdot 1^2$	5	4	0.60	0.80
13	7^2	7	2	1.00	0.29
17	9^2	9	2	1.00	0.22
19	$9^2 \cdot 1^2$	9	4	0.53	0.44
41	$10^4 \cdot 1^3$	10	6	0.70	0.60
97	49^2	49	2	1.00	0.04

Theorem 5 can be proven by direct computation together with Theorem 4, but can also be obtained from Theorem 4 by factoring the action of the permutation A over the subgroup of scalars.

By the way, we can work conversely, starting with the simpler computation for Theorem 5, and then lift projective permutations to linear operators.

Percival and Vivaldi took just this path, but hid all of the simple projective geometry of the situation behind the complicated algebraic theory of field extensions.

Comparing Theorems 4 and 5 with the characteristics of random permutations from Sections 3.2 and 3.3, we can note that the Young diagrams of the automorphisms of finite tori differ greatly both from those of typical random permutations of the same number of elements, and also from the Young diagrams defined by the tables of Galois fields.

To be specific, the value of fullness λ for automorphisms is significantly larger, while the reduction in fullness observed for random permutations (as the area of their Young diagrams increases) is apparently absent in the case of of Fibonacci automorphisms.

The asymmetry μ of the diagrams of automorphisms is also noticeably higher than for random permutations or for permutations generated by tables of Galois fields. The diagrams for automorphisms are more often high ($\mu > 1$ in Theorem 4). At the same time, diagrams of random permutations of large numbers of elements is usually low ($\mu < 1$), but as the area of the diagram increases, the average ratio $\mu = y/x$ of the height of the diagram to its length seems to approach 0.

The significant asymmetry of the Young diagram of automorphisms makes their statistics very different not only from the uniform averaging of the statistics over all $n!$ permutations which we computed in Section 3.1, but also from the statistics of Vershik-Plancherel, whose average value for the ratio y/x is equal to 1. (See Editor's note 3, page 107.)

On the basis of all this we come to the conclusion that the distribution by period of periodic points of an automorphism of the finite torus leads to new asymptotic Young diagrams (as compared with the two others we've studied, they are uniform and Plancherel-like) even for the Fibonacci automorphism with matrix $\begin{pmatrix} 2 & 1 \\ 1 & 1 \end{pmatrix}$.

One might think that the similar phenomena (perhaps even with the same universal asymptotics as $n \to \infty$, or perhaps with dependence of this asymptotics on the continued fractions corresponding to the matrix) occur also for other automorphisms of tori.

It would be interesting to study the behavior as $n \to \infty$ of the average values of the parameters x, y, λ, μ of Young diagrams of cycles for the automorphisms A of a torus of n^2-points M along the whole group of automorphisms (or at least the behavior of their Cesàro average as $n \to \infty$).

It would also be interesting to compare the average over the whole symmetric group $S(n)$ with the average over the subgroup of projective permutations $n = p+1$ of points of a finite projective line $P^1(\mathbb{Z}_p)$.

In the case of automorphisms of a finite m-dimensional torus we must look at the "projective permutations" of the set of $n = p^{m-1} + \cdots +1$ points of a finite $(m-1)$-dimension projective space $P^{m-1}(\mathbb{Z}_p)$. In this case, we may have to distinguish the Lobachevskian part Λ (Klein model) of the projective space and the additional finite relativistic world of De Sitter, $P^{m-1} \setminus \Lambda$.

Finally, the behavior of all these objects as the dimension M approaches infinity can lead to new interesting asymptotics of "enlarged Young diagrams". The point is that for automorphisms of a m-dimensional torus

$$A : \mathbb{Z}_n^m \to \mathbb{Z}_n^m$$

these asymptotic characteristics might depend on a multi-dimensional continued fraction of the operator $A \in GL(n, \mathbb{Z})$. For example, the answers might depend on a "triangulation" by convex integer polyhedra of the continuous torus $(S^1)^{m-1}$, which determines the geometry of the "periods" of the multi-dimensional continued fraction of the operator A.

For the "Young diagram of cycles" in this case we must probably use not just the lengths of the cycles of the operator A, but also a description of the action of the whole commutative group $\mathbb{Z}^{m-1}$ of symmetries of a periodic multi-dimensional continued fraction.

Even the average values of the characteristics of the Young diagram of the operators A, averaged over the entire group of automorphisms (and not even dependent on the continued fraction) requires attentive study (although the study will be empirical, with the aid of experiments, as described above.)

Remark. A permutation of n^m points of a finite torus $\mathbb{Z}_n^m$, given by the matrix $A \in SL(m, \mathbb{Z})$, has a defined period $T(A, n)$, for which

$$A^{T(A,n)} = 1 \quad (\text{mod } n).$$

All the lengths of the cycles of this permutation of n^m elements divide the integer $T(A, n)$.

For this reason we might want to compare the statistics of a Young diagram of cycles of the operator A, even the Fibonacci operator $A = \begin{pmatrix} 2 & 1 \\ 1 & 1 \end{pmatrix}$, as $n \to \infty$ not with the statistics of all the Young diagram of all permutations of n^m elements, but just with the statistics of those permutations the lengths of whose cycles divide a given integer $T(A, n)$.

It is not clear how much this statistic differs from the general statistics of all $n^m!$ permutations of the set $n^m = |\mathbb{Z}_n^m|$ elements. Neither is it clear (although it is doubtless easier to find out) how the period $T(A, n)$ behaves as $n \to \infty$, neither for the Fibonacci matrix $\begin{pmatrix} 2 & 1 \\ 1 & 1 \end{pmatrix}$, nor for a random matrix A in $SL(2, \mathbb{Z}_n)$, nor for a random matrix in $SL(m, \mathbb{Z}_n)$, nor for their projective versions (described above).

All these cases are interesting and accessible to the listeners (at least experimentally). For $n = (2, 3, \ldots, 20)$, Theorem 4 gives us the following periods $T(A, n)$ for the Fibonacci matrix $A = \begin{pmatrix} 2 & 1 \\ 1 & 1 \end{pmatrix}$:

$$T(A, n) = (3, 4, 3, 10, 12, 8, 6, 12, 60, 5, 12, 14, 24, 20, 12, 18, 12, 9, 60).$$

I computed the average values of the characteristics of permutations for the case $n = 5$ over all the permutations of $n^2 = 25$ elements.

The length of each cycle of these permutations divides the number
$$T \begin{pmatrix} 2 & 1 \\ 1 & 1 \end{pmatrix} = 10,$$ and each has exactly one cycle of length one, like the Fibonacci permutations.

The number of such permutations is approximately $25!! \cdot 2^2 \cdot 3^5 \cdot 7 \cdot 11/5$, where $(25)!! = 25 \cdot 23 \cdot 21 \cdot \ldots \cdot 3 \cdot 1$.

The average value of the parameters of these permutations turn out to be as follows:

$$x_T \approx 10.00, \quad y_T \approx 6.96, \quad \lambda_T \approx 0.56, \quad \mu_T \approx 0.70.$$

The values observed for the Fibonacci permutations with $n = 5$

$$x = 10, \quad y = 5, \quad \lambda = 0.50, \quad \mu = 0.50$$

are closer to the values obtained above for permutations whose cycles divide the number T than to the averages over all $25!$ permutations of 25 elements obtained above (see the table in the end of Section 3.2) and equal to

$$\widehat{x} \approx 9, \ \widehat{y} \approx 5, \ \widehat{\lambda} \approx 0.56, \ \widehat{\mu} \approx 0.56.$$

All the same, the observed values differ greatly from the statistics shown above describing the permutations, the length of whose cycles divide the number T.

The behavior of these statistics as $n \to \infty$ remains unknown. I don't even know how the strange function $T \left(\begin{pmatrix} 2 & 1 \\ 1 & 1 \end{pmatrix}, n \right)$ of the number n defining the finite torus of n^2 elements which was permuted above behaves for large values of n. The first ten values of this function given above have not allowed me to give any prediction about its behavior, but listeners may move quickly here, at least in computing the next value of the period T for larger n.

Some readers of the lectures from 2005 have already continued my computations (with the help of their computers). Their computations confirm the correctness of my conjectures about the characteristics of random permutations (see the discussion after Theorem 2). For Fibonacci automorphisms of finite tori they show that the length and height of the diagrams grow on the order of the square root of the number of points on the finite torus with an average fullness of about

0.8 and an average asymmetry which is also on the order of the square root.

For Fibonacci permutations of the finite projective line the Young diagram turn out to be almost rectangular (of the form k^a or $k^a 1^2$ with a even, and equal to 2 in 60% of the cases, according to M. Kazaryan and V. Kleptsyn.)

I hope that the readers will also prove these new conjectures.

See also the article [4].

Editor's notes

1. The "partition function" $n \mapsto p_n$ (which measures the number of partitions of a positive integers n) has been studied for at least three centuries. A lot is known about it. First, calculating its values does not require actual listing all partitions. An enthusiastic high school student armed with a reliable pocket calculator can find the first fifty values in half an hour. How to do that is explained in the book "Mathematical Omnibus" D. Fuchs and S. Tabachnikov [**12***] (see Section 3.4). Here are a few results beyond the table in the present book:

$$p_{20} = 627, \qquad p_{50} = 204226, \qquad p_{100} = 190569292.$$

There are also good asymptotic formulas for p_n. The best known one is due to Hardy and Ramanujan:

$$p_n \sim \frac{1}{4n\sqrt{3}} e^{\pi\sqrt{\frac{2n}{3}}}.$$

By the way, this formula shows that p_n grows faster than any polynomial of n, but slower then any exponential function a^n (with $a > 1$). The Hardy-Ramanujan formula does not give good precision. If you use it to compute p_{10} you will get ≈ 48, and it computes p_{100} and p_{1000} with the errors, respectively, of $\approx 4\%$ and $\approx 1.5\%$. There exist formulas that do not look as good as the Hardy-Ramanujan formula, but which provide much better approximation. (These can be found on the web.)

2. For example, the permutations

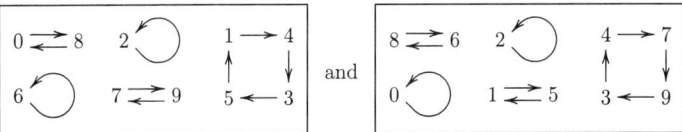

and

are obtained from one another by rearranging digits, or, as the author says, are "indistinguishable from a relativistic point of view". In other words, these two permutations, u and v, are related by a formula $v = tut^{-1}$ where for t one can take the permutation which specifies the rearranging of digits, which in our example is

x	0	1	2	3	4	5	6	7	8	9
$t(x)$	8	4	2	5	7	3	0	1	6	5

By the way, this definition of conjugacy (u is conjugate to v if $v = tut^{-1}$ for some t) can be applied to any group. For the permutation group $S(n)$, it is seen from what was said before that the two permutations are conjugate, if and only if they yield identical partitions of n into the lengths of cycles. Thus, there is a bijection between the conjugacy classes in $S(n)$ and partitions of n.

3. Here the author refers to the famous result concerning the typical limiting shape of the Young diagram, published in 1977 in two independent works. One was written by B. Logan and L. Shepp, and the other by A. Vershik and S. Kerov. The problem solved by these four authors is different from the problem considered in the present book. V. I. Arnold considers all Young diagrams of the area n, each is repeated a number of times equal to the number of permutations with this Young diagram. For example, the seven Young diagrams of area 5,

correspond, respectively, to 1, 10, 15, 20, 20, 30, 24 permutations (the sum of these numbers is $5! = 120$). Thus, the problem considered in the present book, for $n = 5$, is the following: we make $1, 10, 15, \ldots, 24$ copies of the seven Young diagrams shown above, and choose randomly one of the 120 Young diagrams obtained. Then we determine expected characteristics (x, y, λ, μ) of this randomly chosen diagram.

There exists a different way of assigning positive integers to Young diagrams, which is very common in representation theory. A *Young tableau* is a Young diagram of some area n filled with integers $1, \ldots, n$ such that in each row the integers grow to the right, and in each column integers grow down. For example, the Young diagram

has two such fillings:

The number of fillings of the Young diagram λ (which can be computed in several different, but equivalent, ways) is denoted by $\dim V_\lambda$. The reason for this notation (and simultaneously, the explanation of the importance of this number of fillings), comes from the representation theory of the group $S(n)$. An *irreducible representation* of $S(n)$ is a finite-dimensional vector space V with an action of $S(n)$ which does not have any proper invariant subspaces (that is subspaces $W \subset V$, different from 0 and V), which are mapped into themselves by any element of $S(n)$. There is a classical theorem in representation theory that there exists finitely many non-isomorphic irreducible representations of $S(n)$ (p_n of them), and they may be labeled by the Young diagrams of area n. The irreducible representation corresponding to the Young diagram λ is denoted by V_λ, and $\dim V_\lambda$ is the dimension of V_λ. For the seven Young diagrams with $n = 5$ listed above, these dimensions are, respectively, $1, 4, 5, 6, 5, 4, 1$. We assign to a Young diagram λ the "weight" $(\dim V_\lambda)^2$. Thus, for the seven diagrams of area 5, these weights are, respectively, $1, 16, 25, 36, 25, 16, 1$. The sum of these weights is 120, and there is a general theorem stating that the sum $\sum_\lambda (\dim V_\lambda)^2$ is $n!$ for any n. The fraction $(\dim V_\lambda)^2/n!$ is called the *Plancherel measure* of the Young diagram λ. Now, for any fixed n, we make $(\dim V_\lambda)^2$ copies of every Young diagram λ of area n. Thus we obtain $n!$ diagrams. Then we randomly choose one of them and looking at its shape. The theorem of Logan, Shepp, Vershik, and Kerov states that for a large n *almost all* $(99\%, 99.9\%, 99.99\%, \ldots$ — dependingly on n) Young diagrams of area n will have almost the same shape; below we describe this shape. (For example, for $n = 5$, instead of $1, 10, 15, 20, 20, 30, 24$ diagrams, we take $1, 16, 25, 36, 25, 16, 1$ diagrams with the same total of 120.)

Theorem (Logan, Shepp, Vershik, Kerov). *The limiting shape, for $n \to \infty$, of the average Young diagram (with respect to the Plancherel measure) of area n, after rescaling to area 1 and rotation by $135°$, is a curvilinear triangle bounded by the segments of length 2 on the coordinate axes and the so called "Omega curve",*

$$y = \frac{2}{\pi} \left(x \arcsin \frac{x}{2} + \sqrt{4 - x^2} \right),$$

(see the picture below).

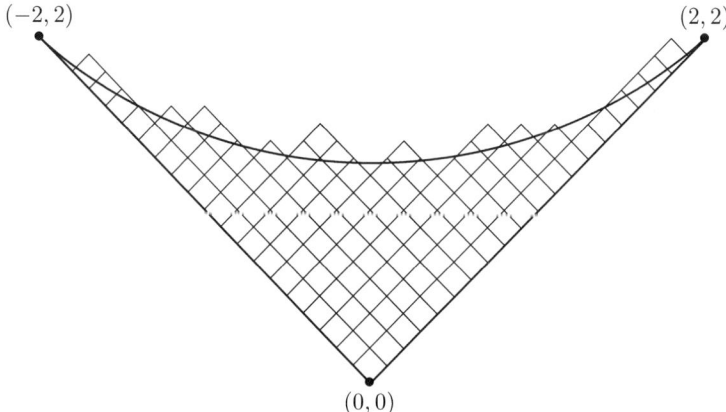

The area of the curvilinear triangle is $1/4$ of the area of the ambient square. Thus, the limit of the fullness λ is $1/4$. To obtain the actual Young diagram, we need to stretch the figure in the picture $\sqrt{n}$ times. The height and the length of this diagram will be, approximately, $2\sqrt{n}$. Thus, the limiting characteristics x, y, λ, μ are $2\sqrt{n}, 2\sqrt{n}, 1/4, 1$, very much different from the ones in Arnold's statistics.

Notice in conclusion, that it is very unlikely that the last theorem can be confirmed experimentally: for values of n for which direct computations are possible, the (Plancherel) average Young diagram has very much different shape.

Lecture 4

The Geometry of Frobenius Numbers for Additive Semigroups

> "Sex is akin to mathematics: in both cases satisfying one's curiosity often leads to pleasure."
>
> *C. Djerassi, "Newton's Darkness"*

> "But if I err, it's no big thing for me,
> I'll not give up my error, all the same."
>
> *B. L. Pasternak, for A. A. Akhmatova*

The subject of this lecture relates to the simplest questions of arithmetic: which numbers can we obtain from a given set of numbers by adding them (in any combinations)?

For example, suppose we have coins in the denominations of 3 cents and 5 cents.

What sums can we make of 3 cent and 5 cent coins?

Clearly we cannot make sums of 1, 2, or 4 cents, and we certainly can make up 3, 5, and 6 cents. But we cannot make up 7 cents. After that we have:

$$8 = 3 + 5, \quad 9 = 3 + 3 + 3, \quad 10 = 5 + 5.$$

And from this it is clear that we can get any larger integer number of cents (by adding 3 cent coins to 8, 9, 10).

It is interesting to draw the set of all attainable sums. It forms an additive semigroup; that is, if the set contains two elements x and y, then it also contains their sum $x + y$. We will also include 0 in the set of sums.

The semigroup with generators 3 and 5 is pictured in Figure 1 (the squares):

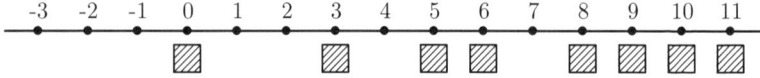

Figure 1. The semigroup generated by 3 and 5 cent coins.

It is interesting to note that the complementary set of integers is situated symmetrically to this semigroup (see the squares in Figure 2).

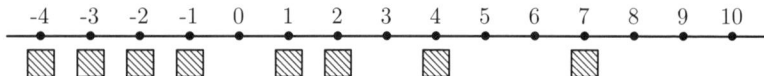

Figure 2. The complement of the semigroup generated by 3 and 5 cent coins.

More precisely, if x is in the semigroup, then $7 - x$ is in its complement. For example, the semigroup contains all the integers greater than or equal to 8, while its complement includes all integers less than or equal to the integer -1, which is complementary to 8.

1. Sylvester's Theorem and the Frobenius Numbers

The first American mathematician, J. Sylvester, proved that the situation will be similar for a semigroup generated by any two positive integers a and b, whose greatest common divisor is 1. This semigroup consists of all possible combinations $xa + yb$ (with non-negative integer coefficients x and y).

Theorem 1 (Sylvester). *The semigroup generated by two relatively prime integers a and b contains all the integers starting with $N(a,b) = (a-1)(b-1)$.*

We will prove theorem 1 below, in Section 4.6.

The symmetry (about the center $(N-1)/2$) also always holds: x *belongs to the semigroup if and only if $y = N - 1 - x$ does not belong to it.*

Frobenius' problem consists in understanding what the situation is in the case of more than two generators. For example, suppose we choose n generators (positive integers)

$$a_1, a_2, \ldots, a_n.$$

If the greatest common divisor of all these numbers is equal to 1, then *the additive semigroup of their integer combinations*

$$\{a = x_1 a_1 + \cdots + x_n a_n\}, \quad x_s \geq 0,$$

contains all the integers greater than or equal to some Frobenius number $N(a_1, \ldots, a_n)$.

This statement is easy to deduce from Sylvester's theorem, by adding generators.

Frobenius' problem consists in actually computing the Frobenius number $N(a_1, \ldots, a_n)$ or at least investigating its behavior when we change the generators a_ϑ (for example, as $a_\vartheta \to \infty$).

Sylvester's formula shows that for $n = 2$ the growth is of the order the product of the generators: $N(a, b) \sim ab$.

For n generators, as we will see below, the role of the product ab starts to be played by the quantity

$$N_0(a_1, \ldots, a_n) = \sqrt[n-1]{(n-1)! a_1 \cdots a_n}.$$

For example, for $n = 3$ the formula is as follows:

$$N_0(a, b, c) = \sqrt{2abc}.$$

Example. The values of $N(a,b,c)$ for $a+b+c=7$ form the following equilateral triangle (in which I have shown the three axes of symmetry):

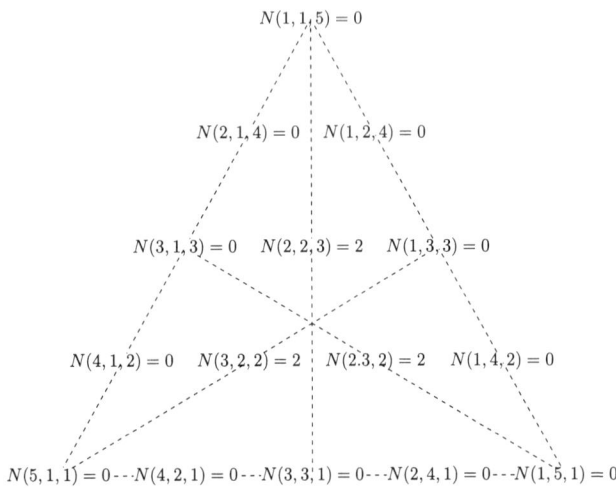

$N(1,1,5) = 0$

$N(2,1,4) = 0 \quad N(1,2,4) = 0$

$N(3,1,3) = 0 \quad N(2,2,3) = 2 \quad N(1,3,3) = 0$

$N(4,1,2) = 0 \quad N(3,2,2) = 2 \quad N(2,3,2) = 2 \quad N(1,4,2) = 0$

$N(5,1,1) = 0\text{---}N(4,2,1) = 0\text{---}N(3,3,1) = 0\text{---}N(2,4,1) = 0\text{---}N(1,5,1) = 0$

Example. For $a+b+c=19$ Frobenius numbers $N(a,b,c)$ form the equilateral triangle shown below.

```
                            0
                          0 : 0
                        0  14  0
                      0   2 : 2   0
                    0  12 24 12  0
                  0   4  6 : 6  4   0
                0  10  8 30  8 10  0
              0   6 18 12:12 18  6   0
            0   8 12 12 32 12 12  8   0
          0   8 14 18 10:10 18 14  8   0
        0   6 12 18 24 30 24 18 12  6   0
      0  10 18 12 10 30 30 10 12 18 10  0
    0   4  8 12 32 10 24 10 32 12  8  4   0
  0  12  6 30 12 12 18:18 12 12 30  6 12  0
0   2 24  6  8 18 12 14 12 18  8  6 24  2   0
0  14  2 12  4 10  6  8 : 8  6 10  4 12  2 14  0
0---0--0--0--0--0--0--0--0--0--0--0--0--0--0--0
```

As we will see below (following the work of V. I. Arnold [**2**]), as the sum $\sigma = a_1 + \cdots + a_n$ grows, these fillings (of a regular n-dimensional simplex) by the Frobenius numbers $N(a_1, \ldots, a_n)$ generate some peculiar asymptotic regularities. We will go on to prove some of these regularities, while others remain regularities in the sense of natural science and await a rigorous mathematical theory (possibly given by the listeners to this lecture).

2. Trees Blocked by Others in a Forest

Suppose a cyclist rides along a straight path (Figure 3) and comes to the vertex of a forest shaped like an angle. Looking at that angle, he first sees the tree at the edge, but then the trees grow in number, and starting at some point N, they become dense (without gaps).

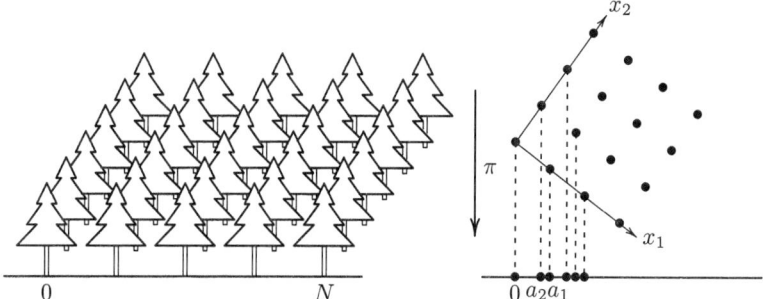

Figure 3. The projection of a forest in the form of a square lattice onto a road.

If the trees are planted in a square lattice, then those points on the road where the cyclist sees a tree, looking only perpendicular to the road, form an additive semigroup (of points along the road, where the point nearest the vertex of the forest is chosen as the origin).

In just this way, the general problem of additive semigroups can be seen as a problem about a linear projection

$$\pi : \mathbb{Z}_+^n \to \mathbb{R}$$

of an n-dimensional quadrant $\{x_1, \ldots, x_n\}$, $x_s \in \mathbb{Z}$. The projection π assigns to a "tree" $x = (x_1, \ldots, x_n) \in \mathbb{Z}_+^n$ an element of the semigroup

$$a = x_1 a_1 + \cdots + x_n a_n.$$

Without worrying for now about mathematical rigor, let us try to estimate the first spot N at which the forest starts to appear dense.

In our model, let us assume that our semigroup $(a_1, \ldots, a_n)$ is generated by natural numbers, so that the whole semigroup $\pi\left(\mathbb{Z}_+^n\right)$ consists of integers.

Let us try to understand how many of them are less than or equal to a fixed number ℓ (Figure 4). Let us call this number $M(\ell)$.

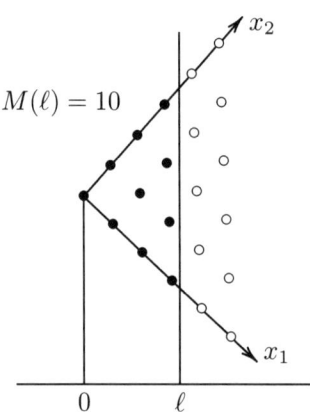

Figure 4. The simplex of initial trees.

We will assume that the area (the n-dimensional volume, for $n > 2$) of one square (of the n-dimensional cube) of the lattice-planted forest is equal to 1. Then the number $M(\ell)$ will be roughly equal to the area (n-dimensional volume) of a triangle (a simplex of dimension n) $S(\ell)$ in $\mathbb{R}^n$, given by the inequalities

$$S(\ell) = \left\{ x \in \mathbb{R}^n : x_s \geq 0, \sum_{s=1}^{n} x_s a_s \leq \ell \right\}.$$

The legs of this right triangle (or simplex) have lengths ℓ/a_s.

Therefore its area is $V(\ell) = \dfrac{1}{2}\dfrac{\ell}{a_1}\dfrac{\ell}{a_2}$. In the n-dimensional case, the n-dimensional volume of the simplex $S(\ell)$ is

$$V(\ell) = \frac{1}{n!} \prod_{s=1}^{n} \frac{\ell}{a_s}.$$

Thus, we arrive at a hypothetical formula approximating the number of trees in a simplex $S(\ell)$:

$$M(\ell) \approx \frac{\ell^n}{n!\,\Pi}, \text{ where } \Pi = \prod_{s=1}^{n} a_s.$$

For the projection π to cover point ℓ, the inequality $M(\ell) - M(\ell-1) \geq 1$ must hold.

If we think of ℓ as very large and substitute the derivative for the difference we obtain the condition $\dfrac{dM}{d\ell} \geq 1$, that is:

$$\frac{\ell^{n-1}}{(n-1)!\,\Pi} \geq 1,$$

or

$$\ell \geq N_0 = \sqrt[n-1]{(n-1)!\,\Pi}. \tag{1}$$

For $n = 2$ we get the approximate condition $\ell \geq a_1 a_2$ (asymptotically close to Sylvester's exact condition $\ell \geq (a_1 - 1)(a_2 - 1)$).

The problem of a rigorous mathematical foundation for formula (1) is not at all simple, and we will discuss it below, in Sections 4.3 and 4.4.

3. The Geometry of Numbers

We will begin our attempt to put formula (1) on a mathematical basis with a very simple idea (due to Minkowski), connecting the number of integers $M(\ell)$ with the volume $V(\ell)$: exactly in what sense is $M(\ell) \approx V(\ell)$? We denote the sum $a_1 + \cdots + a_n$ by σ.

Theorem 2. *The following inequality holds:*

$$V(\ell) \leq M(\ell) \leq V(\ell + \sigma).$$

Proof. With each integer point x of the closed region $S(\ell)$ we associate a unit n-dimensional cube (Figure 5):

$$\{y \in \mathbb{R}^n : x_s \leq y_s \leq x_s + 1, s = 1, \ldots, n\}.$$

The union of these cubes, constructed for all integer points x of a closed simplex $S(\ell)$ forms a polyhedron P. This polyhedron is contained in the closed simplex $S(\lambda + \sigma)$.

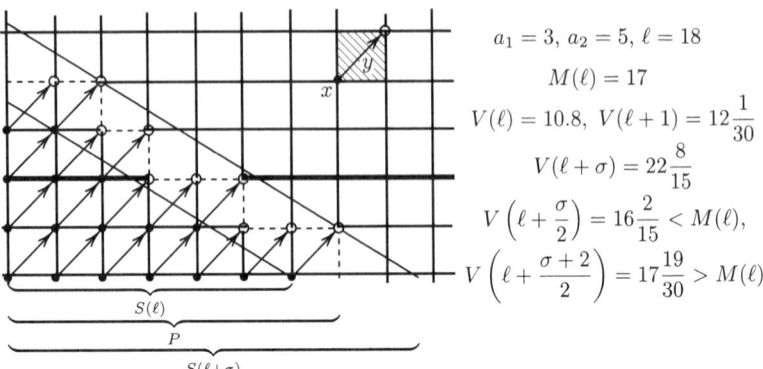

$$a_1 = 3, \ a_2 = 5, \ \ell = 18$$

$$M(\ell) = 17$$

$$V(\ell) = 10.8, \ V(\ell + 1) = 12\frac{1}{30}$$

$$V(\ell + \sigma) = 22\frac{8}{15}$$

$$V\left(\ell + \frac{\sigma}{2}\right) = 16\frac{2}{15} < M(\ell),$$

$$V\left(\ell + \frac{\sigma + 2}{2}\right) = 17\frac{19}{30} > M(\ell)$$

Figure 5. The circumscribed polyhedron P of the simplex $S(\ell)$.

Indeed, for any point $y = (y_1, \ldots, y_n)$ of the cube associated with point $x = (x_1, \ldots, x_n) \in S(\ell)$, we find:

$$\sum y_s a_s \leq \sum (x_s + 1) a_s = \sum x_s a_s + \sum a_s \leq \ell + \sigma.$$

The polyhedron P contains the closed simplex $S(\ell)$.

Indeed, we can replace each coordinate x_s of point x in $S(\ell)$ with its integer part z_s. The point z thus obtained also lies in the closed simplex $S(\ell)$, and therefore the original point x belongs to the cube constructed on the integer point z. This means that x belongs to the polyhedron P.

So we now have the relation

$$S(\ell) \subseteq P \subseteq S(\ell + \sigma). \qquad \square$$

Remark. Experimentation shows that a more interesting inequality holds involving the number $M(\ell)$ of integer points in a closed simplex $S(\ell)$:

$$V\left(\ell + \frac{\sigma - 2}{2}\right) \leq M(\ell) \leq V\left(\ell + \frac{\sigma + 2}{2}\right).$$

The arithmetic mean of the left and right estimates gives a particularly good approximation to the number $M(\ell)$.

Example. For the triple $\{a_s\} = \{3, 5, 8\}$ it is easy to compute these values. In this case $\sigma/2 = 8$.

l	0	1	2	3	4	5	6	7	8	9	10
$V(l)$	1	1	1	2	2	3	4	4	6	7	8
$M(l)$	0	$\dfrac{1}{720}$	$\dfrac{1}{90}$	$\dfrac{3}{80}$	$\dfrac{4}{45}$	$\dfrac{25}{144}$	$\dfrac{3}{10}$	$\dfrac{343}{720}$	$\dfrac{32}{45}$	$\dfrac{81}{80}$	$\dfrac{25}{18}$

l	11	12	13	14	15	16	17	18	19	20	21
$V(l)$	10	11	13	15	17	20	22	25	28	31	35
$M(l)$	1.84	2.4	3.05	3.81	4.69	5.69	7.60	8.10	9.52	11.11	12.86

The cases of even and odd sums σ differ a bit, so I present one more table, for the triple $\{a_s\} = \{3,5,7\}, \sigma = 15, (\sigma+1)/2 = 8$.

l	0	1	2	3	4	5	6	7	8	9
$V(l)$	1	1	1	2	2	3	4	5	6	7
$M(l)$	0	$\dfrac{1}{630}$	$\dfrac{8}{630}$	$\dfrac{3}{70}$	0.102	0.198	0.343	0.544	0.812	1.157

l	10	11	12	13	14	15	16	17	18	19	20
$V(l)$	9	10	12	14	16	19	21	24	27	30	34
$M(l)$	1.59	2.11	2.74	3.49	4.35	5.36	6.50	7.79	9.25	10.88	12.69

Here the inequalities

$$V\left(\ell + \frac{\sigma - 1}{2}\right) = V(\ell + 7) \leq M(\ell) \leq V(\ell + 8) = V\left(\ell + \frac{\sigma + 1}{2}\right).$$

are satisfied for $\ell \geq 3$, but for $\ell = 2$ the situation is different:

$$\left(V\left(7\frac{1}{2}\right) \approx 0.984\right) < (m(2) = 1) < V(8).$$

I don't know a general proof of the inequalities given above, even for the case $n = 2$ of planar polygons (where the idea of symmetry with respect to the points of the line $(a, x) = \sum a_s x_s = \ell + \sigma/2$ inspires hope).

Now we will use Theorem 2, proven above, to give a lower estimate of the Frobenius numbers.

Consider the interval $N \leq \ell < N + r$, where N is the Frobenius number. For each point ℓ of this interval there exists an integer point $x(x_s \geq 0)$, in which $(a, x) = \ell$. Therefore in the simplex $S(N + r - 1)$ there are at least r integer points:

$$M(N + r - 1) \geq r.$$

Using the upper bound for the number of integer points provided by Theorem 2, we now obtain a lower bound for the volume of the corresponding simplex. This gives us a lower bound for the Frobenius number N.

From Theorem 2 we conclude that

$$r \leq M(N + r - 1) \leq V(N + \sigma + r - 1) = \frac{(N + \sigma + r - 1)^n}{n!\Pi}.$$

For the value $r = \lambda(N + \sigma - 1)$ we find

$$\lambda(N + \sigma - 1) \leq \frac{(N + \sigma - 1)^n (1 + \lambda)^n}{n!\Pi},$$

whence

$$N + \sigma - 1 \geq \left(\frac{\lambda n!\Pi}{(1 + \lambda)^n}\right)^{\frac{1}{n-1}}.$$

Thus

$$N \geq \omega\Pi^{\frac{1}{n-1}} - \sigma + 1,$$

where the coefficient ω has the following form:

$$\omega(\lambda) = \left(\frac{\lambda n!}{(1 + \lambda)^n}\right)^{\frac{1}{n-1}}.$$

For $\lambda = \dfrac{1}{n - 1}$ we have:

$$\left(\frac{\lambda}{(1 + \lambda)^n}\right)^{\frac{1}{n-1}} = \frac{n - 1}{n^{1 + \frac{1}{n-1}}} \geq \frac{1}{4},$$

so that our estimate for the Frobenius number has the form

$$N \geq \frac{1}{4}(n!\Pi)^{\frac{1}{n-1}} - \sigma + 1.$$

This inequality indicates that the growth is of the order $\Pi^{1/(n-1)}$, that is, $\sigma^{n/n-1}$, as was given by our heuristic reasoning in Section 4.2, although the coefficient in the estimate we just proved is smaller for $n = 2$ than in Sylvester's formula ($N_0 = \Pi - \sigma + 1$).

4. Upper Bound Estimate of the Frobenius Number

Our estimate will use simple general facts of the geometry of numbers.

In the Euclidean space $\mathbb{R}^n$, we consider the basis $(P_1, P_2, \ldots, P_n)$ and the lattice $\mathbb{Z}^n \subset \mathbb{R}^n$ that they generate. We will now look for an upper bound for the largest possible radius R of a ball in this space, which doesn't contain any points of the lattice.

Theorem 3. *The inequality*

$$R \leq \frac{\sqrt{R_1^2 + R_2^2 + \cdots + R_n^2}}{2}$$

holds, where R_s is the distance from a point P_s to the space generated by the vectors $P_1, P_2, \ldots, P_{s-1}$.

Proof. For $n = 1$ this is obvious: $R \leq |P_1|/2$. Suppose that the assertion is already proven for $n = k$.

In the space $\mathbb{R}^{k+1}$ generated by the vectors $P_1, P_2, \ldots, P_{k+1}$, we consider the hyperplane Q generated by the vectors $P_1, P_2, \ldots, P_k$, and the parallel hyperplanes Q_j passing through points jP_{k+1} respectively. (See Figure 6.) The center of an empty ball in $\mathbb{R}^{k+1}$ is contained in one of the layers of thickness R_{k+1} between the hyperplanes Q_j and so is at a distance of at most $R_{k+1}/2$ from one of them.

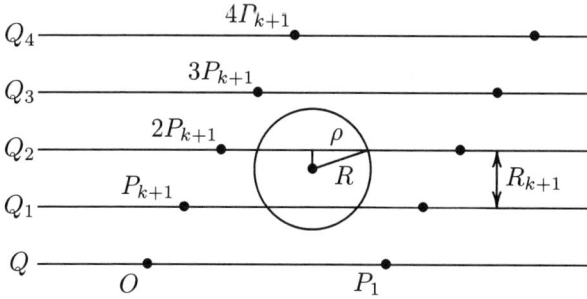

Figure 6. An empty ball in the space with a lattice of dimension $k + 1$.

The intersection of our empty ball with this hyperplane Q_j is empty in this hyperplane, and so its radius ρ is approximated by the value suggested for $n = k$ in Theorem 3:

$$\rho \leq \frac{\sqrt{R_1^2 + \cdots + R_k^2}}{2}.$$

From the Pythagorean Theorem we have

$$R^2 \leq \rho^2 + (R_{k+1}/2)^2 = (R_1^2 + \cdots + R_k^2 + R_{k+1}^2)/4,$$

which proves Theorem 3. $\square$

We will now apply this theorem to the lattice of integer points on the plane

$$\{x \in \mathbb{R}^n : (a, x) = \ell\},$$

in which there are no integer points in the orthant $x_s \geq 0$. We will denote by $S(\ell)$ the $(n-1)$-dimensional simplex of intersection of this hyperplane with the orthant. We will divide the dependence of functions on the vector $a = (a_1, \ldots, a_s)$ into its dependence on its direction $\alpha(a) = a/\sigma(a)$ and on its "size" $\sigma(a)$.

Theorem 4. *The radius of a sphere inscribed in the simplex $S(\ell)$ is equal to $\beta(\alpha)\ell/\sigma(a)$, where the dimensionless coefficient $\beta(\alpha)$ is*

$$\beta(\alpha) = \frac{|\alpha|}{\sum_{s=1}^{n}(\alpha_s \sqrt{|\alpha|^2 - \alpha_s^2})},$$

which depends only on the direction of the vector a.

Example. For $n = 2$ this formula assumes the form

$$\beta(\alpha) = \frac{|\alpha|}{2\Pi} = \frac{\sqrt{\alpha_1^2 + \alpha_2^2}}{2\alpha_1\alpha_2} \geq \sqrt{2}.$$

For any n, we have $\beta(\alpha) \geq \sqrt{n/n-1} \geq 1$.

Proof of Theorem 4. A pyramid with base $S(\ell)$ has a volume

$$V(\ell) = \frac{1}{n!} \prod_{s=q}^{n} \left(\frac{\ell}{a_s}\right) = \frac{\ell^n}{n!\Pi}.$$

This volume can also be expressed in terms of the volume $|S(\ell)|$ of the base and the length h of the altitude dropped on it from O:

$$V(\ell) = \frac{1}{n}h|S(\ell)|, \quad h = \frac{\ell}{|a|}.$$

Therefore

$$|S(\ell)| = \frac{nV(\ell)}{h} = \frac{\ell^{n-1}|a|}{(n-1)!\Pi} = \frac{\ell^{n-1}\sigma(a)|\alpha|}{(n-1)!\Pi}. \tag{1}$$

On the other hand, the volume of the simplex $S(\ell)$ is equal to $\frac{1}{n-1}$ times the product of the radius r of the sphere inscribed in it and the sum of the areas $|S_s|$ of its faces $\{x_s = 0\}$. Therefore

$$r = \frac{n-1}{\sum_{s=1}^{n} S_s}|S(\ell)|. \tag{2}$$

Using formula (1) to compute the areas of the faces we find

$$S_s = \frac{\ell^{n-3}}{(n-2)!}\frac{a_s\sqrt{|a^2| - a_s^2}}{\Pi} = \frac{\sigma^2(a)\ell^{n-2}\alpha_s\sqrt{|a^2| - \alpha_s^2}}{(n-2)!\Pi}. \tag{3}$$

Substituting expressions (1) and (3) into formula (2) we find:

$$r = \frac{(n-1)\ell^{n-1}\sigma(a)|\alpha|}{(n-1)!\Pi} \cdot \frac{(n-2)!\Pi}{\sigma^2(a)\ell^{n-2}\sum_{s=1}^{n}(\alpha_s\sqrt{|\alpha|^2 - \alpha_s^2})}$$

$$= \frac{\ell}{\sigma(a)}\frac{|\alpha|}{\sum_{s=1}^{n}(\alpha_s\sqrt{|\alpha|^2 - \alpha_s^2})},$$

which proves Theorem 4. $\qquad\square$

We now note that if there are no integer points in the simplex $S(\ell)$, then there are also none in the inscribed ball. Therefore the radius of this inscribed ball cannot be larger than the bounds given by Theorem 3 (applied to the $(n-1)$-dimensional hyperplane $\{x \in \mathbb{R}^n : (a, x) = \ell\}$). This theorem gives us the inequality

$$\frac{\beta(\alpha)\ell}{\sigma(a)} \leq \frac{\sqrt{R_1^2 + \cdots + R_{n-1}^2}}{2}, \quad \ell \leq \frac{\sigma(a)}{2\beta(a)}\sqrt{R_1^2 + \cdots + R_{n-1}^2}. \tag{4}$$

Theorem 5. *For the vector* $a = (a_1, \ldots, a_n) \in \mathbb{R}^n$ *with any direction* $\alpha = a/\sigma(a)$ *we have the following upper bound for for the Frobenius number:*

$$N(a) \leq 1 + \gamma(\alpha)\sigma^2(a)$$

(where γ *is a constant that will be defined in the proof).*

Proof of Theorem 5. Consider the "flag" of subspaces $\mathbb{R}^1 \subset \mathbb{R}^2 \subset \cdots \subset \mathbb{R}^n$, where $\mathbb{R}^s$ is spanned by the first s coordinate axes (Figure 7).

Let us take the hyperplane W^{n-1} in $\mathbb{R}^n$, given by the equation $a_1 x_1 + \cdots + a_n x_n = 0$. In each space R^s, we introduce a vector A_s with components $a_1, \ldots, a_s$, and define the hyperplane W^{s-1} orthogonal to it by the equation

$$(A_s, X_s) = 0 \text{ (where } X_s = (x_1, \ldots, x_s) \in \mathbb{R}^s).$$

Thus, we have defined a flag of vector subspaces in the hyperplane W^{n-1}:

$$0 \subset W^1 \subset \cdots \subset W^{n-1}$$

(for example, W^1 is a line on the plane $\mathbb{R}^2$, given by the equation $(A_2, X_2) = 0$; that is, by the equation $a_1 x_1 + a_2 x_2 = 0$).

The intersection of the hyperplane W^s with the integer lattice $\mathbb{Z}^{s+1}$ of the space $\mathbb{R}^{s+1}$ defines an s-dimensional lattice Γ_s in this s-dimensional hyperplane.

Lemma. *The volume of a fundamental s-dimensional parallelepiped Δ_s of the lattice Γ_s in the Euclidean space W^s is equal to* $|\Delta_s| = |A_{s+1}|/d_s$ *where d_s is the greatest common divisor of the numbers $a_1, \ldots, a_{s+1}$ and $|A_{s+1}|^2 = a_1^2 + \cdots + a_{s+1}^2$, so that $|A_{s+1}|$ is the Euclidean length of the vector A_{s+1}.*

Proof of Lemma. We can obtain a fundamental parallelepiped of a lattice $\mathbb{Z}^{s+1}$ in the space $\mathbb{R}^{s+1}$ by taking a fundamental parallelepiped Δ_s of the lattice Γ_s in W^s lying on the hyperplane W^s and adding to it the integer vector $x \in \mathbb{Z}^{s+1} \setminus W^s$ nearest to W^s (Figure 8).

The scalar product of this vector with the vector A_{s+1} normal to the hyperplane W^s gives us the smallest possible positive value for

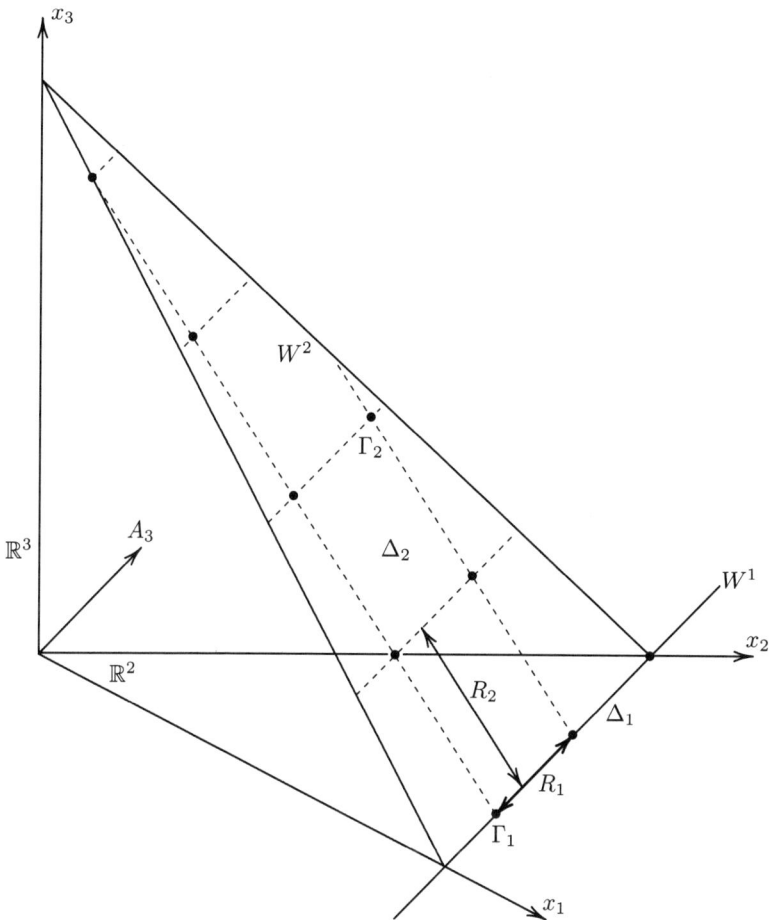

Figure 7. Consecutive altitudes R_s and parallelepipeds Δ_s for $n = 3$. Instead of the plane $(A_3, x) = 0$, we have drawn the plane $(A_3, x) = \text{const}$, where the lattice Γ_2 is the same.

the integer linear combination

$$a_1 x_1 + \cdots + a_{s+1} x_{s+1} = (A_{s+1}, x).$$

The value of this linear combination is divisible by the greatest common divisor d_s of its coefficients, and its smallest value is in fact d_s.

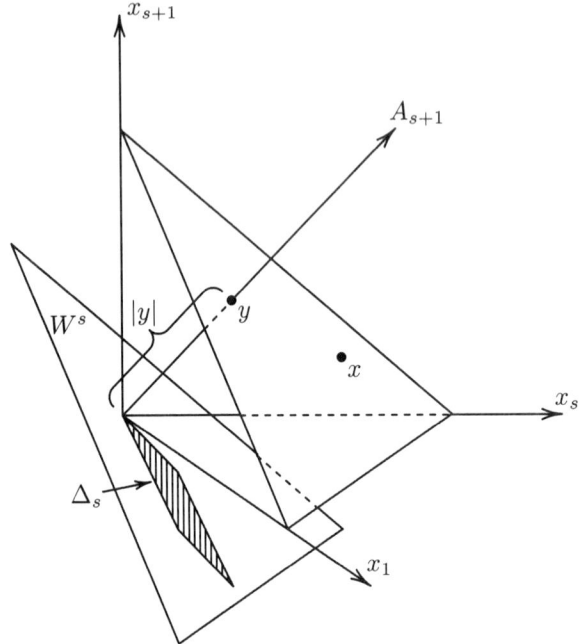

Figure 8. Computation of the volume $|\Delta_s|$ of a fundamental parallelepiped Δ_s of the lattice Γ_s in W^s

Thus the integer vector x nearest to the hyperplane W^s but not lying on it satisfies the relationship $(A_{s+1}, x) = d_s$ (Figure 8).

Let us now find the distance from this nearest vector to the hyperplane W^s.

To this end, we draw the normal to the hyperplane

$$\left\{ y = \frac{|y| A_{s+1}}{|A_{s+1}|} \right\}.$$

Point y of this normal lies at the same distance $\rho = |y|$ from the hyperplane W^s as the nearest integer point x, if $(A_{s+1}, y) = (A_{s+1}, x) = d_s$.

Taking the scalar product of y and the vector A_{s+1}, we find that

$$(A_{s+1}, y) = |y|(A_{s+1}, A_{s+1})/|A_{s+1}| = |y||A_{s+1}| = d_s,$$

from which we have $\rho = |y| = d_s/|A_{s+1}|$.

Thus the volume of a fundamental parallelepiped of the lattice $\mathbb{Z}^{s+1}$ in the Euclidean space $\mathbb{R}^{s+1}$ is $|\Delta_s||y| = \rho|\Delta_s|$. But this volume is equal to 1, since $\mathbb{Z}^{s+1}$ is a standard integer lattice.

Hence $|\Delta_s| = 1/\rho = |A_{s+1}|/d_s$, and Lemma is proved.   □

Corollary. *The length R_s of the altitude of a fundamental parallelepiped Δ_s of a lattice Γ_s in the Euclidean space W^s whose base is a fundamental parallelepiped Δ_{s-1} of the sublattice Γ_{s-1} is equal to*

$$R_s = \frac{|\Delta_s|}{|\Delta_{s-1}|}.$$

Proof of Corollary. The s-dimensional volume $|\Delta_s|$ of the parallelepiped Δ_s is equal to the product of the $(s-1)$-dimensional volume $|\Delta_{s-1}|$ of the parallelepiped which is its base and the length R_s of the altitude (Figure 7).

It will be more convenient for us to write this formula for R_s of each altitude in the form

(1) $$R_s^2 = \frac{(a_1^2 + \cdots + a_{s+1}^2)d_{s-1}^2}{(a_1^2 + \cdots + a_s^2)d_s^2}.$$

Note that d_{s-1} is divisible by d_s (since the greatest common divisor of a set of numbers is a divisor of any subset of this set).

Thus the product of the sequence of integers $q_s = \frac{d_{s-1}}{d_s}$ is

$$q_1 q_2 \cdots q_{n-1} = d_0 = a_1,$$

while the product of the lengths of the altitudes has the form

$$R_1^2 \cdots R_{n-1}^2 = \frac{|\Delta_1|^2}{|\Delta_0|^2}\frac{|\Delta_2|^2}{|\Delta_1|^2} \cdots \frac{|\Delta_{n-1}|^2}{|\Delta_{n-2}|^2} = \frac{|\Delta_{n-1}|^2}{|\Delta_0|^2},$$

However, we know the boundary conditions $|\Delta_0| = |A_1|/d_1 = a_1/a_1 = 1$ and $|\Delta_{n-1}|^2 = |A_n|^2/d_n^2 = |A_n|^2/1$. From these we conclude that the product of the lengths of all $n-1$ altitudes is given by the wonderful formula

$$R_1^2 \cdots R_{n-1}^2 = a_1^2 + \cdots + a_n^2. \tag{2}$$

We will use this equality to find an upper bound for the lengths of the altitudes. For this, we first estimate them from below.

From formula (1) for the lengths of the altitudes, it follows that $R_s^2 > 1$ (since $|A_{s+1}|^2 > |A_s|^2$ and $q_s \geq 1$).

Therefore formula (2) for the product of the lengths of the altitudes gives us an upper bound for the lengths of the altitudes:

$$R_s^2 = \frac{|A_n|^2}{R_1^2 \cdots R_{s-1}^2 R_{s+1}^2 \cdots R_{n-1}^2} < |A_n|^2.$$

This proves the following inequality:

$$\sum_{s=1}^{n-1} R_s^2 < (n-1)(a_1^2 + \cdots + a_n^2).$$

This upper bound for the squares of the lengths of the altitudes gives rise, by Theorem 3 of Section 4.4, to an upper bound for the radius r of an empty ball (which has no points in common with the lattice $\Gamma_{n-1} = \mathbb{Z}^n \cap W^{n-1}$) in the hyperplane W^{n-1}:

$$r^2 \leq \frac{n-1}{4}|A_n|^2. \tag{3}$$

By Theorem 4, there exists an empty ball of radius $\dfrac{\ell}{\sigma(a)}\beta(\alpha)$ in the hyperplane W^{n-1} (with the constant β depending only on the direction α of vector a), if $\ell = N(a) - 1$. (Since there are no integer points in the simplex $A_n(x) = \ell$, $x_j \geq 0$, neither are there any in the inscribed ball.)

From inequality (3) we find the following upper bound for ℓ:

$$\frac{\ell^2}{\sigma^2(a)}\beta^2(\alpha) \leq \frac{n-1}{4}|A_n|^2,$$

whence

$$N(a) = \ell + 1 \leq 1 + \frac{\sqrt{n-1}}{2\beta(\alpha)}|A_n|\sigma(a).$$

But $|A_n|^2 = a_1^2 + \cdots + a_n^2 = \sigma^2(a)(\alpha_1^2 + \cdots + \alpha_n^2)$, so that we get an upper bound for the Frobenius numbers:

$$N(a) \leq 1 + \frac{\sqrt{n-1}}{2\beta(\alpha)}\sigma^2(a)\sqrt{\alpha_1^2 + \cdots + \alpha_n^2},$$

which proves Theorem 5 (with the constant $\gamma(\alpha) = \dfrac{\sqrt{n-1}\,|\alpha|}{2\beta(\alpha)}$). $\quad\square$

By the way: sometimes the following estimate, which gives a smaller bound to the Frobenius numbers, is more useful

$$N \leq 1 + \frac{\sigma(a)}{\beta(\alpha)} \sqrt{\left(\sum_{s=1}^{n-1} R_s^2\right)} \Big/ 2 \qquad (4)$$

with expression (1) for R_s^2. For example, when all $d_s = 1$, the length of each altitude R_s depends only on the direction α of the vector a, except for $R_1^2 = a_1^2 + a_2^2$.

Remark. The lower bound for the Frobenius number N (Section 4.3) had the form $N \geq \text{const}\,(\alpha)\Pi^{\frac{1}{n-1}} \left(\geq \text{const}\,(\alpha)\sigma^{1+\frac{1}{n-1}}\right)$. As σ increases, this quantity grows more slowly then σ^2 for $n > 2$. For example, for $n = 3$, the lower bound turns out to be $N \geq \text{const}\,\sigma^{3/2}$, while the upper bound is $N \leq \text{const}\,\sigma^2$, which is much greater.

The following example illustrates the inescapable nature of this phenomenon: we will examine examples where $N(a, b, c)$ is certainly greater than or equal to a constant times σ^2 (which, for large σ is as many times larger than $\sigma^{3/2}$ as you like), since an upper bound for a quantity that grows as $\sigma^{3/2}$ does not exist for $n = 3$.

Example. Let us look at three (relatively prime) numbers

$$a, b, c = a + b.$$

In this case the Frobenius number is easy to compute:

$$N(a, b, c) = N(a, b)$$

(since the sum of copies of the numbers (a, b, c) are just sums of copies of the numbers a and b).

So $N(a, b, c) = (a-1)(b-1)$ (by Sylvester's formula). We will need only the quadratic growth of σ proven above even without Sylvester's formula (which, however, is proven below, in Section 4.6).

Let us now suppose that $1/9 < a/b < 9$, while $a > 2$, $b > 2$ (Figure 9). Then $a - 1 > a/2$, $b - 1 > b/2$, and so the inequality

$$N(a, b, c) = (a - 1)(b - 1) > \frac{ab}{4}$$

is satisfied.

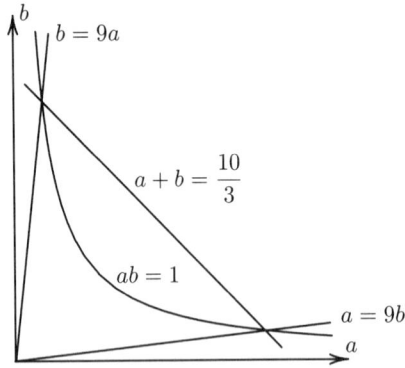

Figure 9. A sector within which $(a+b)^2 < \left(\dfrac{10}{3}\right)^2 ab$.

But $ab > (9/100)(a+b)^2$ in the sector $1/9 < a/b < 9$. Indeed, on the hyperbolic segment $ab = 1$ that is in our sector, the function $a+b$ achieves its maximum at the endpoints of the segment, where $a = 3, b = 1/3$ or $a = 1/3, b = 3$, so that

$$a + b \le 3 + \frac{1}{3} = \frac{10}{3}$$

everywhere in this sector. From homogeneity, this means that the inequality $(a+b)^2 < (10/3)^2\, ab$ is also satisfied for any value of ab in this sector. Therefore the following inequality is satisfied everywhere in our sector:

$$\frac{ab}{4} > \frac{9(a+b)^2}{400},$$

that is,

$$N(a,b,c) > \frac{9}{400}\sigma^2.$$

In particular, the ratio

$$\frac{N(a,b,c)}{\sigma^{3/2}} > \frac{9}{400}\sqrt{\sigma}$$

takes on arbitrarily large values in the given region (with a and b relatively prime). Hence an upper bound of the form

$$N(a,b,c) < \operatorname{const}(\alpha)\sigma^{3/2}$$

is impossible (for many triples in a whole sector of directions α).

Thus there are examples of quadratic growth of Frobenius numbers with respect to $\sigma(\alpha)$. Nevertheless, I was not able to find out, *how numerous* triples with this property of quadratic growth of the Frobenius number with respect to σ are: are they typical or exceptional?

The point is that arguments used in our proof of the inequality $N \leq \mathrm{const}\,\sigma^2$ actually prove more (see inequality (4) above). The quantity σ^2, appears rather than $\sigma^{1+\frac{1}{n-1}}$ because we estimated the growth of the lengths of all the altitudes R_s by $|A_n|^2$. However, we know only that this is the rate of growth of their product

$$R_1^2 \cdots R_{n-1}^2 = |A_n|^2,$$

so that the lengths of the altitudes themselves grow slower.

In the case when all the *ratios* of the lengths of the altitudes R_s/R_t are bounded from above by some constant (not dependent on $|A_n|$), we could have deduced from the expression given in (2) for the product the bounds for the altitudes of the form $R_s^2 \leq \mathrm{const}\,|A_n|^{\frac{2}{n-1}}$. This is much smaller, than the value $|A_n|^2$ that we have used.

For example, consider the case when the coefficients $a_1, a_2, \ldots, a_n$ can be ordered in such a way that all the greatest common divisors of the first of them,

$$d_1 = (a_1, a_2), \ d_2 = (a_1, a_2, a_3), \ldots, d_{n-2} = (a_1, a_2, \ldots, a_{n-1}),$$

are equal to 1, so that formula (1) takes the form

$$R_s^2 = \frac{a_1^2 + \cdots + a_{s+1}^2}{a_1^2 + \cdots + a_s^2} = \frac{\alpha_1^2 + \cdots + \alpha_{s+1}^2}{\alpha_1^2 + \cdots + \alpha_s^2} = \Psi_s(\alpha),$$

which is independent of $\sigma(a)$. In this case the condition for boundedness of the ratio R_s/R_t of the lengths of the altitudes defines a sector in the space of directions, within which the bound $N \leq 1 + \mathrm{const}\,\sigma^{1+\frac{1}{n-1}}$ holds. So the asymptotic behavior of the Frobenius number N of order $\sigma^{\frac{n}{n-1}}$ (that is, $\Pi^{\frac{1}{n-1}}$) holds in some directions, while the bound of order σ^2, (that is, $\Pi^{2/n}$) holds in other directions. And the question of which directions prevail is not simple.

Even the question of the behavior of the average $N(\sigma)$ of the numbers $N(a)$ as it depends on σ in every direction of the vector

a, with a sum $\sigma(a)$ for its coordinates, is not simple and deserves experimental investigation.

The numerical data given below indicate a rate of growth of the average Frobenius numbers less than σ^2, but it might be slower for large values of the sum $\sigma = a + b + c$. (I only got up to $\sigma = 41$, but see the description of the further results in the end of next section.)

5. Average Values of the Frobenius Numbers

In order to compare the bounds we've proven for the Frobenius numbers with real examples I computed directly the values of all the Frobenius numbers $N(a, b, c)$ for $\sigma = 41$; that is, for all 780 triples of natural numbers whose sum $a + b + c$ is 41.

For the value of the sum σ I chose a prime number in order to exclude cases of "commensurability", in which all the numbers a_s have a nontrivial common divisor, so that the additive semigroup they generate won't contain all the natural numbers starting with N for any value of N.

The triples $N(a, b, c)$ of numbers with a given sum σ naturally form an equilateral triangle. For $\sigma = 7$ or 19, these triangles are given above in Section 4.1. For $\sigma = 41$ it is sufficient to draw just a part of this entire triangle (whose 6 symmetries then easily allow us to recover the missing parts). Nonetheless, we present the whole triangle on the next page.

It is clear from the tables of numbers $N(a, b, c)$ given above (and below) that the ratio

$$\vartheta = \frac{N(a, b, c)}{(2abc)^{1/2}}$$

can change significantly with even a small change in the arguments a, b, c.

Example $(a + b + c = 41)$.

$N(7, 14, 20) = 114$, significantly larger than $\sqrt{2 \cdot 7 \cdot 14 \cdot 20} \approx 62.4$.

$N(7, 15, 19) = 47$, significantly smaller than $\sqrt{2 \cdot 7 \cdot 15 \cdot 19} \approx 63.2$.

```
                                    0
                                  0   0
                                0  36  0
                              0   2   2   0
                            0  34  68  34  0
                          0   4   6   6   4   0
                        0  32   8  96   8  32   0
                      0   6  62  12  12  62   6   0
                    0  30  12  34 120  34  12  30   0
                  0   8  14  18  20  20  18  14   8   0
                0  28  56  84  24 140  24  84  56  28   0
              0  10  18  24  28  30  30  38  34  18  10   0
            0  26  20  34  32  38 156  38  32  34  20  26   0
          0  12  50  30 100  56  28  28  56 100  30  50  12   0
        0  24  24  72  40  40  48 168  48  40  40  72  24  24   0
      0  14  26  36  44  50  54  56  56  54  50  44  36  26  14   0
    0  22  44  34  48 110  39  46 176  46  39 110  48  34  44  22   0
  0  16  30  34  32  34  40  70  44  44  70  40  34  32  34  30  16   0
0  20  32  60  80  44  72  68  47 180  47  68  72  44  80  60  32  20   0
0  18  38  48  60  50 114  52  56  90  90  56  52 114  50  60  48  38  18  0
0  18  36  34  34  46  47  54  44  48 180  48  44  54  47  46  34  34  36  18  0
0  20  38  34  68  80  48  44 104  56  62  62  56 104  44  48  80  68  34  38  20  0
0  16  32  48  34  80  96 112  56  64  54 176  54  64  56 112  96  80  34  48  32  16  0
0  22  30  60  60  46  48 112 120  70  52  60  60  52  70 120  48  46  60  60  30  22  0
0  14  44  34  80  50  47  44  56  70 140  62 168  62 140  70  56  44  47  50  80  34  44  14  0
0  24  26  34  32  44 114  54 104  64  52  62 156 156  62  52  64 104  54 114  44  32  34  26  24  0
0  12  24  36  48  34  72  52  44  56  54  60 168 156 168  60  54  56  44  52  72  34  48  36  24  12  0
0  26  50  72  44 110  40  68  56  48  62 176  60  62  62  60 176  62  48  56  68  40 110  44  72  50  26  0
0  10  20  30  40  50  39  70  47  90 180  62  54  52 140  52  54  62 180  90  47  70  39  50  40  30  20  10  0
0  28  18  34 100  40  54  46  44 180  90  48  56  64  70  70  64  56  48  90 180  44  46  54  40 100  34  18  28  0
0   8  56  24  32  56  48  56 176  44  47  56  44 104  56 120  56 104  44  56  47  44 176  56  48  56  32  24  56   8  0
0  30  14  84  28  38  28 168  56  46  70  68  52  54  44 112 112  44  54  52  68  70  46  56 168  28  38  28  84  14  30  0
0   6  12  18  24  30 156  28  48  54  39  40  72 114  47  48  96  48  47 114  72  40  39  54  48  28 156  30  24  18  12   6  0
0  32  62  34  20 140  30  38  56  40  50 110  34  44  50  46  80  80  46  50  44  34 110  50  40  56  38  30 140  20  34  62  32  0
0   4   8  12 120  20  24  28  32 100  40  44  48  32  80  60  34  68  34  60  80  32  48  44  40 100  32  28  24  20 120  12   8   4  0
0  34   6  96  12  34  18  84  24  34  30  72  36  34  34  60  48  34  34  48  60  34  34  36  72  30  34  24  84  18  34  12  96   6  34  0
0   2  68   6   8  62  12  14  56  18  20  50  24  26  44  30  32  38  36  38  32  30  44  26  24  50  20  18  56  14  12  62   8   6  68   2  0
0  36   2  34   4  32   6  30   8  28  10  26  12  24  14  22  16  20  18  18  20  16  22  14  24  12  26  10  28   8  30   6  32   4  34   2  36  0
0   0   0   0   0   0   0   0   0   0   0   0   0   0   0   0   0   0   0   0   0   0   0   0   0   0   0   0   0   0   0   0   0   0   0   0   0   0   0   0
```

In the first case, $\theta \approx 1.82$, while in the second, $\vartheta \approx 0.74$, which is twice as small, although the triples are neighbors.

It has been a while since I put forth the conjecture that the empirically natural "asymptotics" $N \sim \left((n-1)! \prod_{s=1}^{n} a_s\right)^{\frac{1}{n-1}}$ must be considered a weak approximation. That is, the closeness to it emerges (as $\sigma(a)$ grows) only for average values N of the vector a, whose directions α vary over some region in the space of directions. (See V. I. Arnold [2]).

This conjecture remains unproven (see Editor's note 1, page 154), but I decided to verify at least the behavior of the average value of the functions N and $I = \sqrt{abc}$ along the simplex $\{a + b + c = \sigma,\ a \geq 1,\ b \geq 1,\ c \geq 1\}$.

These computations, for the table given above with $\sigma = 7, 19$ and 41, lead to the following average values $\widehat{N}$ and $\widehat{I}$.[1]

σ	$\sum N$	$\sum 1$	$\widehat{N} = \dfrac{(\sum N)}{(\sum 1)}$	$\sum I$	$\widehat{I} = \dfrac{(\sum I)}{(\sum 1)}$
7	6	15	0.4	43.04	2.87
19	1332	153	8.7	1880	12.29
41	33126	780	42.47	31068	39.83
97	909930	4560	199.546		
199	12975216	19503	665.293		

We can use logarithms with these data to get the order of the presupposed power asymptotics:

$$\widehat{N} \sim C\sigma^u.$$

Indeed, if $\ln \widehat{N} = \ln C + u \ln \sigma$, then the coefficient u can be found as the slope of the graph of $\widehat{N}$, drawn on log-log graph paper:

$$u \approx \frac{\ln \widehat{N}(\sigma_2) - \ln \widehat{N}(\sigma_1)}{\ln \sigma_2 - \ln \sigma_1}.$$

For $\sigma_1 = 7$, $\sigma_2 = 41$ and for $\sigma_1 = 19$, $\sigma_2 = 41$ we have

$$u \approx \frac{3.83 + 0.92}{3.70 - 1.95} \approx 2.5 \quad \text{and} \quad u \approx \frac{3.83 - 2.20}{3.70 - 2.94} \approx 2.1.$$

Choosing $(\sigma_1 = 41, \sigma_2 = 97)$ gives a slope $u = 1.8$ for the line, while the choice $(\sigma_1 = 97, \sigma_2 = 199)$ gives us $u \approx 1.6$. These numbers, which are close to 1.5, partly support my conjecture of 1999, stating that for large vectors a the Frobenius numbers $N(a)$ grow on the average as σ to the power $1 + 1/(n - 1)$; that is, as σ to the power $3/2$, if $n = 3$. The computations for $\sigma = 97$ and 199 were done by computer by A. Goder, at my request.

Analogous computations for $\widehat{I}$ lead even more quickly to $u \approx 1.5$, which is the value given by the asymptotics proven above (and by similar considerations, applied to the integral of the function $I = \sqrt{abc}$ over the simplex $a + b + c = \sigma$).

---

[1] The symbol $\sum 1$ means the number of triples (a, b, c) satisfying the conditions formulated above. [transl.]

6. Proof of Sylvester's Theorem

Here we consider integer points ($x \geq 0$, $y \geq 0$) on the line $ax + by = \ell$ in the plane $\mathbb{R}^2$ (where a and b are relatively prime natural numbers, and ℓ is an integer).

Theorem 6. *If $\ell = N-1 = ab-a-b$, then there are no integer points on segment of the real line considered, but if $\ell \geq N = (a-1)(b-1)$, then there are such points.*

Proof. First we look at all the integer points on the line $ax + by = 0$. Since x and y are relatively prime, the number x must be divisible by b, and the number y by a. Therefore the non-zero integer point P nearest to 0 on our line has coordinates $x = b, y = -a$ (Figure 10).

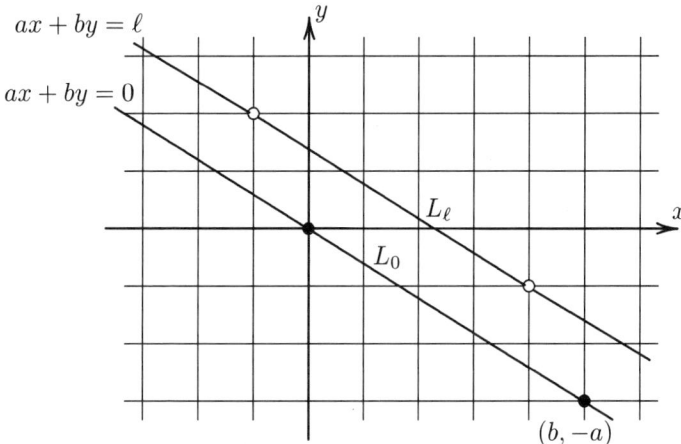

Figure 10. Integer points on the lines $ax + by = \ell$ ($a = 3$, $b = 5$, $\ell = 7$).

It follows that the distance between neighboring integer points on our line is $L = \sqrt{a^2 + b^2}$.

On every line $ax + by = \ell$ parallel to this one (where ℓ is an integer) the integer points form a lattice with the same step L ((Figure 10), since these lines are taken into each other by a shift of the plane which preserves the lattice of integer points. (This is because the equation $ax + by = 1$ has solutions in integers, as is clear for example from the

Euclidean algorithm for finding the greatest common divisor 1 of the numbers a and b).

From all this it follows that for $\ell \geq ab$ there exist integer points on the segment of the line $ax + by = \ell$, with $x \geq 0$, $y \geq 0$, (since the length of this segment is $\sqrt{\left(\dfrac{\ell}{a}\right)^2 + \left(\dfrac{\ell}{b}\right)^2} = \dfrac{\ell}{ab}\sqrt{a^2 + b^2} \geq L$).

Thus, the second assertion of the theorem is proved for $\ell \geq ab$.

Let us now consider the line $ax + by = ab$ (Figure 11).

The points $A(x = b, y = 0)$ and $B(x = 0, y = a)$ lie on this line.

The points $A'(x = b - 1, y = -1)$ and $B'(x = -1, y = a - 1)$ lie on the line $ax + by = \ell'$, where $\ell' = ab - a - b = N - 1$.

The distances $|AB|$ and $|A'B'|$ are equal to L. Therefore there are no integer points on the segment $A'B'$ of the line $ax + by = \ell'$, except for the endpoints A' and B'.

This completes the proof of the first assertion of the theorem (about the lack of integer points on the interval of the line with $\ell = N - 1$).

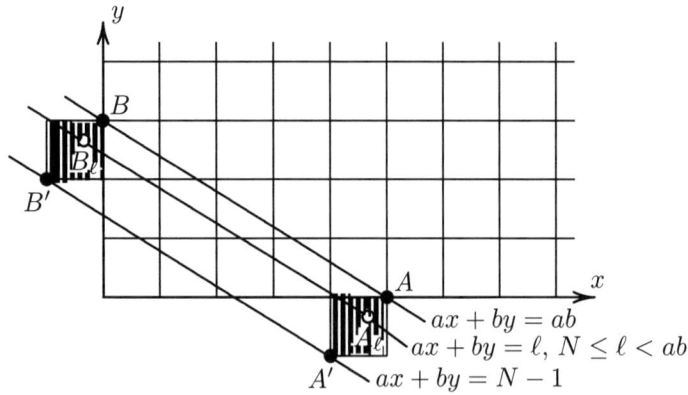

$ax + by = ab$
$ax + by = \ell, \ N \leq \ell < ab$
$ax + by = N - 1$

Figure 11. Sylvester's lines $ax + by = \ell, N - 1 \leq \ell \leq ab$

For a full proof of the second assertion of the theorem it remains to verify the assertion only for $N \leq \ell < ab$. Under these conditions, the line $ax + by = \ell$ intersects the diagonals (AA') and (BB') drawn in the shaded squares (Figure 11) respectively at the points A_ℓ and B_ℓ, and it is clear that A_ℓ has a negative y-coordinate, while B_ℓ has a negative x-coordinate.

The distance $|A_\ell B_\ell|$ is equal to L, just as is the distance $|AB|$. Therefore there must be an integer point (x_ℓ, y_ℓ) on segment $A_\ell B_\ell$. Both coordinates of this integer point are non-negative, since the intersections of segment $A_\ell B_\ell$ of the line $ax + by = \ell$ with the regions $x < 0$ and $y < 0$ lie inside the shaded squares, where there are no integer points. The presence of an integer point (x_ℓ, y_ℓ), where $x_i \geq 0$, $y_i \geq 0$, $ax_i + by_i = \ell \geq N$, completes the proof of Sylvester's theorem. □

7. The Geometry of Continued Fractions of Frobenius Numbers

I thought up the geometric theorems given below about the Frobenius numbers in 1999, when I wrote an article about weak asymptotics of the number of solutions of Diophantine problems and computed thousands of Frobenius numbers $N(a, b, c)$. But I didn't publish these results, considering them too obvious. That they are obvious is shown below. In contrast to their proofs, the discovery of these geometric facts is not at all simple.

Let (a, b, c) be positive integers with no common divisor greater than 1. We consider the function $\ell(y, z) = by + cz$ with values on the closed positive quadrant $\{y \geq 0, z \geq 0\}$.

Definition 1. We call an integer point r of the quadrant a *realizer* of the remainder k upon division by a, if the value $\ell(r)$, gives a remainder of k when divided by a and the value of $\ell(r)$ is minimal (among all points on the closed positive quadrant $\{(y \geq 0, z \geq 0)\}$ which give a remainder of k upon division by a).

Definition 2. We call the set of all realizers of all a remainders $(k = 0, 1, \ldots, a - 1)$ the *domain* $D(a, b, c)$ of the triple (a, b, c). It is more convenient not to draw the finite set of realizers, but rather the corresponding real domain where they lie (see Theorem 7).

All the remainders are realized, if $(a, b, c) = 1$: this follows from the Euclidean algorithm. As a rule, there is only one realizer for every remainder, in which case the number of realizers is equal to a. But if there are more realizers, it will not hinder us.

Theorem 7. *The broken line bounding the domain D is always a 'staircase' (a Young diagram): if a point q lies outside the domain D, then any point $Q \geq q$ (with coordinates $\{(y(Q) \geq y(q),\ z(Q) \geq z(q)\})$ also lies outside the domain D.*

Example. For $a = 21, b = 31, c = 45$, if we write down the remainder of the number $\ell(a)$ upon division by a next to every point $q = (y, z)$, we get a collection of remainders, giving us the realizers indicated in boldface, forming a domain D bounded by a staircase line. (See Figure 12)

Consider a vertex M of the staircase which surrounds region D where the value L of the linear function ℓ is maximal (there may be several such "maximal vertices").

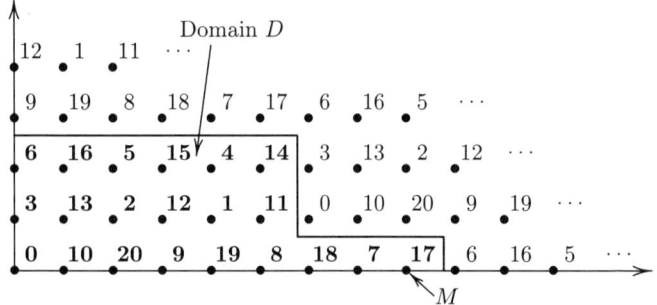

Figure 12. The staircase for the triple $(21, 31, 45)$.

Theorem 8. *The Frobenius number $N(a, b, c)$ is given by the formula*

$$N = L - a + 1.$$

Example. In the preceding example, the (unique) maximal vertex M has coordinates $y = 8,\ z = 0$, so that

$$L = \ell(8, 0) = 8 \cdot 31 + 0 \cdot 45 = 248 \quad (\equiv 17 \pmod{21}).$$

In this case, Theorem 8 asserts that $N(21, 31, 45) = 248 - 21 + 1 = 228$, which is true, but not very easy to show.

Proof of Theorem 7. If point q lies outside the domain D, then there is a realizer r in D with a corresponding value $\ell(r)$ which is congruent to $\ell(q)$ modulo a and such that $\ell(r) < \ell(q)$. Let $Q \geq q$. Let us prove that the point Q also lies outside the domain D.

Consider the vector $R = r + (Q - q)$. We have $\ell(R) \equiv \ell(Q)$ (mod a), $\ell(R) < \ell(Q)$, and so point Q is not a realizer and lies outside the domain D. This proves Theorem 7. $\qquad\square$

Proof of Theorem 8.

Lemma 1. *The number $L - a$ does not belong to the additive semigroup $P = \{ax + by + cz, \ x \geq 0, \ y \geq 0, \ z \geq 0\}$ of linear combinations of (x, y, z) with non-negative integer coefficients.*

Proof of Lemma 1. If we could have a representation

$$L - a = aX + bY + cZ, \ X \geq 0, Y \geq 0, Z \geq 0, \tag{1}$$

then point Q with coordinates (Y, Z) would satisfy the conditions

$$\ell(Q) = bY + cZ = L - a - aX \equiv (L = \ell(M)) \bmod a$$

$$(\ell(Q) = bY + cZ) \leq (aX + bY + cZ = L - a) < (L = \ell(M)).$$

That would mean that point M would not be a realizer, which contradicts its definition. This contradiction proves that representation (1) is not possible. Lemma 1 is proved. $\qquad\square$

Lemma 2. *Any integer $K > L - a$ lies in the additive semigroup P.*

Proof of Lemma 2. The realizer r of the remainder of $K > L - a$ when divided by a belongs to the domain D of the triple (a, b, c). Therefore the inequality $\ell(r) \leq L < K + a$ is satisfied (since the number L is the maximum of the function ℓ over all the realizers).

The remainders upon division by a of K and $\ell(r)$ are identical. Hence the inequality $\ell(r) < K + a$ implies the inequality $\ell(r) \leq K$. Therefore $K = \ell(r) + xa$, where the integer x is non-negative. If (y, z) are the coordinates of point r, we have the equality

$$K = ax + by + cz,$$

which proves Lemma 2. $\qquad\square$

Theorem 8 is implied by Lemmas 1 and 2: the Frobenius number N is larger than $L - a$ by Lemma 1, and is not greater than $L - a + 1$ by Lemma 2, so is equal to $L - a + 1$. $\qquad\square$

Example. In the case when $a = b$ we find that point M with coordinates $(y = 0, z = a - 1)$ is the maximal realizer, so $L = c(a - 1)$ and Theorem 8 gives us the answer $N(a, a, c) = c(a - 1) - a + 1 = (a - 1)(c - 1)$. This gives us a new proof of Sylvester's theorem: $N(a, c) = (a - 1)(c - 1)$.

Theorem 9. *If a point Q lies outside the domain D, one unit higher than a horizontal segment of the boundary (where $z = $ const), then the number $\ell(Q)$ is congruent modulo a to the value $\ell(R)$ at some point R of the lower boundary of the domain D (where $z = 0$).*

Moreover, any realizer R of a remainder of the number $\ell(Q)$ upon division by a lies on the boundary of the domain D.

Proof of Theorem 9. If the realizer R had satisfied the inequality $z(R) > 0$, then the point $r = (y(R), z(R) - 1)$ would have also belonged to the domain D. Then the point $q = (y(Q), z(Q) - 1)$ would be such that the remainder of $\ell(q)$ upon division by a would be the same as the remainder for some smaller value $\ell(r)$.

Therefore point q would lie outside the domain D (it could not be a realizer because of the existence of its rival r). In other words, if $z(R)$ were positive, then the original point Q could not lie immediately above a horizontal segment of the border of the domain D, which contradicts the condition in the statement of the Theorem. Thus $z(R) = 0$, and Theorem 9 is proven. $\qquad\square$

Remark. If y and z change places, we obtain a theorem analogous to Theorem 9 which describes the *vertical* parts of the boundary of D (where $y = $ const): $\ell(Q) = \ell(R = (0, z(R))$ (mod a), on the vertical segment next to the boundary vertical segment.

Corollary. *For any inward vertex of the boundary staircase (that is, a vertex whose spike is directed toward the origin) the value of the function ℓ at a point V outside of D closest to this vertex (which is at a distance 1 from the horizontal as well as from the vertical segments of the staircase) is divisible by a.*

Proof of the Corollary. The realizer r of the remainder of the number $\ell(V)$ upon division by a must belong to the x-axis, to the y-axis, and to the z-axis (by Theorem 9). This means that $r = 0$ and $\ell(r) = 0$. Therefore that $\ell(V)$ is divisible by a. $\qquad\square$

Example. In Figure 13, the point V closest to an inward vertex has coordinates $y(V) = 6, z(V) = 1$. At this point the indicated remainder is 0, since the value $\ell(V) = 6 \cdot 31 + 1 \cdot 45 = 231 = 21 \cdot 11$, which is divisible by $a = 21$.

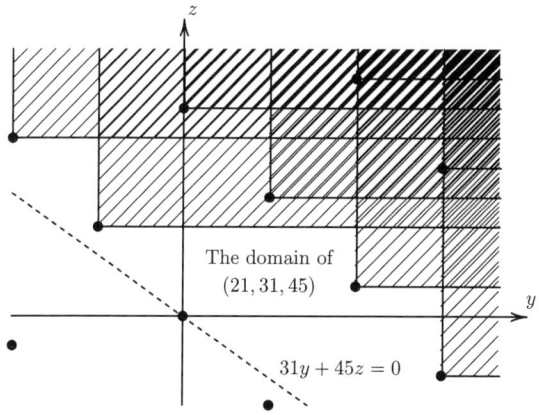

Figure 13. The construction of the saw Π for the example of Figure 12.

Theorems 7–9 provide a quick way to construct the domain D (and therefore to compute the Frobenius numbers).

Namely, we look at the lattice Γ of those integer points q in the plane with coordinates (y, z) for which the value $\ell(q)$ is divisible by a. For example, the points $(c, -b)$ and $(a, 0)$ are among these. (But they don't necessarily form a basis of the lattice Γ.)

We consider that part Γ_+ of the lattice Γ (a "semi-lattice") where the function $\ell = by + cz$ takes on positive values. For each point q of the semi-lattice Γ_+ we construct the closed "tooth"

$$K_q = \{Q \geq q : y(Q) \geq y(q), z(Q) \geq z(q)\}$$

with vertex q. Consider (Figure 13) the "saw" of all such teeth

$$\Pi = \bigcup_{\ell(q) > 0} K_q.$$

Theorem 10. *The domain D is the complement of the saw Π in the positive quadrant $\mathbb{R}^2_+ = \{y \geq 0, \ z \geq 0\}$ (in the plane with coordinates y and z):*

$$D = \mathbb{R}^2_+ \setminus \Pi.$$

Proof of Theorem 10.

Lemma 1. *No quadrant K_q intersects the domain D.*

Proof of Lemma 1. Indeed, if point q lies in the positive quadrant of the plane, then the (unique) realizer of the remainder of $\ell(q)$ when divided by a is the point $0 \in D$. Therefore $q \neq 0$ does not belong to the domain D.

If $y(q) > 0, z(q) < 0$, then we consider the decomposition

$$q = q' - q'', q' = (y(q), 0), q'' = (0, -z(q)).$$

In this notation

$$\ell(q') = \ell(q) + \ell(q'') > \ell(q''),$$

and therefore point q' does not belong to the domain D (because at point q'' the remainder of a upon division by ℓ is the same, but the value is smaller).

By Theorem 7, the whole quadrant of points greater than q' does not intersect the domain D. Therefore the whole quadrant K_q of points greater than q also does not intersect the domain D (since $z \geq 0$ everywhere in D).

Thus $D \cap K_q = \emptyset$ in the cases considered as well, so Lemma 1 is proven. □

Lemma 2. *If a point $Q \in \mathbb{R}^2_+$ does not belong to any of the quadrants K_q (where $\ell(q) > 0$), then point Q belongs to the domain D.*

Proof of Lemma 2. Consider the realizer $R \in D$ of the remainder of the number $\ell(Q)$ upon division by a (Figure 14). If $\ell(R) < \ell(Q)$, then at the point $q = Q - R$ the following conditions are satisfied:

the difference $\ell(q) = \ell(Q) - \ell(R) > 0$ is divisible by a, $Q \geq q$ (that is, $y(Q) \geq y(q)$, $z(Q) \geq z(q)$).

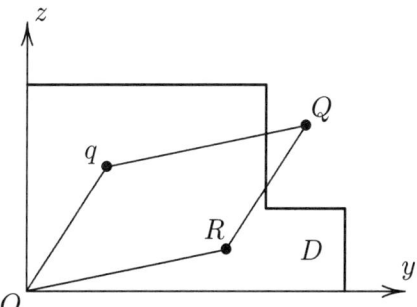

Figure 14. Proof of Lemma 2.

This means that $Q \in K_q$ for the points q of the semilattice, where $\ell(q) > 0$ (point q does not necessarily lie in D).

It follows that if point Q does not lie in any of the quadrants K_q of points of the semilattice, then $\ell(R) = \ell(Q)$. That is, point Q is itself a realizer and belongs to the domain D.

Lemma 2 is proved. $\qquad\square$

Theorem 10 is a direct consequence of Lemmas 1 and 2, so it is now proved. $\qquad\square$

The results proven above about the structure of the domain $D(a, b, c)$ of realizers can be reformulated in terms of the theory of continued fractions in the following way.

For a triple (a, b, c) of integers, we consider the plane $\Pi = \{(x, y, z) : ax + by + cz = 0\}$ in $\mathbb{R}^3$.

Three lines (X, where $x = 0$; Y, where $y = 0$; Z, where $z = 0$) divide the plane Π into six "Weyl chambers" .

The integer points of plane Π form a two-dimensional lattice Γ. For example, it always includes the "Koszul points"

$$\pm(x = 0, \ y = c, \ z = -b), \ \pm(x = -c, \ y = 0, \ z = a),$$

$$\pm(x = b, \ y = -a, \ z = 0),$$

but the basis vectors for Γ maybe different.

The points of the lattice Γ other than O in each (closed) Weyl chamber K form an additive semigroup within it. The convex hull of this semigroup is bounded by a broken line (convex toward zero) which is called a *continued fraction* (of the triple (a, b, c) in the chamber K).

These six continued fractions form a star-shaped hexagon on plane Π with vertices at points of the lattice Γ, inside of which lies only one point O of Γ (Figure 15).

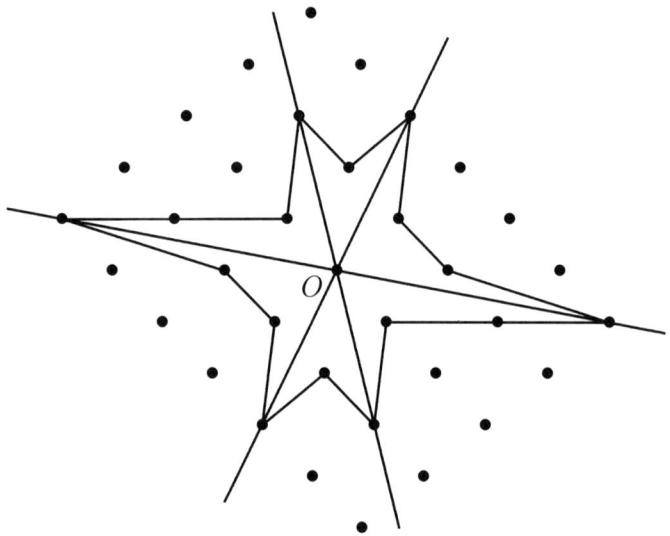

Figure 15. Hexagon of continued fractions of a lattice with three lines.

We will now describe the domain of realizers $D(a, b, c)$ in terms of the geometry of these continued fractions (so that the Frobenius number N of the triple (a, b, c) will also be described in terms of continued fractions).

Remark. One might hope that these geometric constructions would allow in future for a generalization of the Frobenius number to the case of incommensurable arguments (a, b, c) (just as the theory of ordinary continued fractions allows us to generalize the Euclidean algorithm for finding the greatest common divisor of two integers, passing from finite convex broken lines to infinite ones).

So we begin with three lines passing through a point O of the plane $\mathbb{R}^2$ with some lattice Γ (with origin O). We take one of the six angles into which these three lines divide the plane.

Let us call this angle K, and its two sides (Y, Z). Let the third angle be X (Figure 16). Let Γ_+ be the semilattice of the lattice Γ consisting of the points of Γ lying strictly on the same side of line X as angle K. Let us translate angle K, parallel to itself, to every point q of the positive semilattice Γ_+, to get angle K_q with vertex $q \in \Gamma_+$.

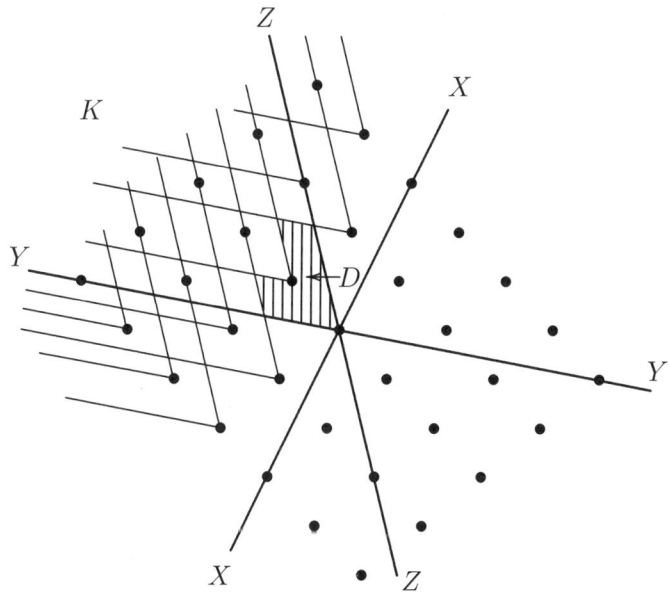

Figure 16. Construction of the domain D for a lattice Γ, angle K, and lines X, Y, and Z.

Definition. We define the *domain D of angle K* (in the lattice Γ and for the triple of lines X, Y, Z) to be the complement of angle K with respect to the union of all the translated angles K_q with vertices $q \in \Gamma_+$:

$$D = K \setminus \bigcup_{q \in \Gamma_+} K_q.$$

Let a, b, c be three positive integers which have no common divisor greater than 1.

Theorem 11. *Suppose we project the three-dimensional space with coordinates (x, y, z) onto the plane with coordinates (y, z) (along the x-axis). Consider the domain D of angle K (where $y \geq 0$, $z \geq 0$) on the plane $\Pi = \{(x, y, z) : ax + by + cz = 0\}$ with the lattice Γ of integer points and the triple of lines*

$$(X : x = 0, \ Y : y = 0, \ Z : z = 0).$$

This domain then projects onto the domain $D(a, b, c)$ of realizers $q = (y, z)$ of remainders upon division by a of the values of the linear function $\ell(y, z) = by + cz$.

Remark. In particular, it follows from this theorem that the Frobenius numbers of all six continued fractions of a Weyl chamber on plane Π coincide. This is not at all obvious geometrically: in fact $N(a, b, c) = N(b, c, a) = \cdots$.

Proof of Theorem 11. We can write the equation $ax + by + cz = 0$ of plane Π in the form

$$x = -\frac{b}{a}y - \frac{c}{a}z;$$

that is, in the form $x = -\ell(y, z)/a$.

From this formula, it follows that the points of lattice Γ (the integer points on plane Π) project exactly onto those integer points q of the plane with coordinates (y, z) for which the value of $\ell(q)$ is divisible by a.

The description of the staircase border of domain $D(a, b, c)$ on the plane with coordinates (y, z) which is given in Theorems 7–10 provides, in terms of projections (from the three-dimensional space onto the plane with coordinates (y, z) along the x-axis), exactly the geometric description shown above of domain D for angle K (where $y \geq 0$, $z \geq 0$) on plane Π (together with the lattice Γ of integer points and the three lines (X, Y, Z)). This proves Theorem 11. $\qquad \square$

Remark. Aside from the theorem proven above, my long-completed computations of thousands of Frobenius numbers also lead to hundreds of other observations, which have not yet found a mathematical explanation or general formulation. Here are a few of these strange experimental observations:

$$\frac{N(13,32,52)=372}{N(13,33,51)=186}=2, \qquad \frac{N(9,43,45)=336}{N(9,42,46)=168}=2,$$

$$\frac{N(5,35,57)=224}{N(5,34,58)=112}=2, \qquad \frac{N(4,20,73)=216}{N(4,19,74)=54}=4,$$

$$N(4,6+4k,87-4k)=90 \quad (k=0,1,2,\ldots,14, k\neq 8),$$

$$N(9,9k\pm 3,88-(9k\pm 3))=168 \quad (k=0,1,2,\ldots,7).$$

It would be interesting to understand how the additive semigroup and the continued fractions of various members of these series of triples of numbers are related, for triples whose Frobenius numbers are related by the formulas shown above.

Extending the theory described above to the case of Frobenius numbers $N(a_1, a_2, \ldots, a_n)$, where $n > 3$, doesn't change the theory by much, except that the continued fractions become multi-dimensional.

A continuation of the investigation of the asymptotic behavior of the Frobenius numbers of large three-dimensional vectors, which is contained in the 2005 lecture in Dubna, is described in the article: V.I. Arnold, [**5**].

These results, confirming the self-similarity of the average distribution of Frobenius numbers about an n-dimensional vector space, show that the growth of Frobenius numbers on the order of σ to the power $1+1/(n+1)$ is more common than the growth on the order of σ^2 (which are observed in resonance locations).

In particular, the average value of the Frobenius numbers $N(a,b,c)$ on the triangle $a+b+c=\sigma$ probably grows as $\sigma^{3/2}$ for large values of σ.

The details of this research will be published in a large article: V.I. Arnold, [**10**].

(See Editor's note 2, page 154.)

8. The Distribution of Points of an Additive Semigroup on the Segment Preceding the Frobenius Number

Sylvester proved that *the points of an additive semigroup with two relatively prime generators ($P = \{xa + yb\}, x \geq 0, \; y \geq 0$) fill up exactly half of the integer segment $\{0, N-1\}^2$ less than the Frobenius number $N(a,b)$. (Specifically, there are $N/2$ of them, since point p lies in the semigroup P if and only if the conjugate point $q = N-1-p$ does not belong to the semigroup P).*

If there are more than two generators, then *the semigroup P will cover no more than half the segment $\{0, N - 1\}$.* Indeed, if point p lies in P, then the conjugate point $q = N - 1 - p$ cannot lie in P (otherwise the sum $p + q = N - 1$ would be in the semigroup P, which contradicts the minimality condition in the definition of the Frobenius number N).

In some cases, the semigroup with three generators occupies exactly half the integer segment $\{0, N - 1\}$.

Example. $N(3, 4, 7) = N(3, 4) = 6$, since the third generator 7 does not add anything to the semigroup. Of the six points $\{0, \ldots, 5\}$, the semigroup contains only three: $\{0, 3, 4\}$.

In some other cases the part of the segment below the Frobenius number is less than the half of this segment.

Example. $N(4, 5, 7) = 7$, but only three points $\{0, 4, 5\}$ of the seven points of the integer segment $\{0, 1, \ldots, 6\}$ belong to the semigroup $P\{4, 5, 7\}$, and $3/7 < 1/2$.

It seems that the semigroup $P(a, b, c)$ always covers no less than a third of the points of the integer segment $\{0, 1, \ldots, N(a, b, c) - 1\}$ (and the lower bound for the domain it covers may possibly be even larger than $1/3$). But this has not been proved.

Attempts to understand why the semigroup cannot occupy too small a part of the segment $\{0, 1, \ldots, N - 1\}$ have led to wonderful

---

[2]The number N of points on this integer segment is even, since $N = (a-1)(b-1)$ would be odd only if both generators were even, which is impossible since they are relatively prime.

experimental observations. I tell about them here because I hope that that the students of the Dubna school will participate in the proof (or refutation) of the striking conjectures formulated below.

Let $p \in \{0, 1, \ldots, N - 1\}$ where $N = N(a, b, c)$ is the Frobenius number for the three generators a, b, c of the additive semigroup $P = \{xa + yb + zc \colon x \geq 0, y \geq 0, z \geq 0\}$ $(x, y, z \in \mathbb{Z}_+)$ which do not have common divisor greater than 1.

Definition 1. A number p is called a $(+)$-*number* if it is an element of the semigroup P.

Example. The number 0 is a $(+)$-number, but the number $N - 1$ is not.

Definition 2. The number p is called a $(-)$-*number* if it is not an element of the semigroup P.

Example. The number $N - 1$ is a $(-)$-number, but the number a is not.

Definition 3. The number q is called *conjugate* to the number p if $p + q = N - 1$. We write $q = \bar{p}$.

Clearly, the conjugate of the number conjugate to p is p itself.

If p is a $(+)$-number, then its conjugate $\bar{p}$ is a $(-)$-number. Otherwise, the number $p + \bar{p} = N - 1$ would be in the semigroup P, which contradicts the definition of the Frobenius number N.

Definition 4. The number p is called a $(-, -)$-number if both p and its conjugate are $(-)$-numbers:

$$p \notin P, \ \bar{p} \notin P.$$

Our next goal will be to investigate the set of $(-, -)$-numbers and the number $\#\{(-, -)\}$ of its elements.

Remark. The numbers $\#\{(+)\}$, $\#\{(-)\}$, and $\#\{(-, -)\}$ of $(+)$-numbers, $(-)$-numbers, and $(-, -)$-numbers are connected by the following obvious relations:

$$\#\{(+)\} + \#\{(-)\} = N,$$
$$\#\{(-)\} - \#\{(+)\} = \#\{(-, -)\}.$$

The second follows from the equalities

$$\#\{(+)\} = \#\{(+,-)\} = \#\{(-,+)\},$$
$$\#\{(-)\} = \#\{(-,-)\} + \#\{(-,+)\}$$

where $\#\{(\alpha,\beta)\}$ is the number of points p of class α for which the conjugate point $\bar{p}$ belongs to class β).

Hence,

$$N = 2\#\{(+)\} + \#\{(-,-)\},$$

and therefore *the Frobenius number N and the number $\#\{(-,-)\}$ have the same parity.*

In particular, *if the Frobenius number N is odd, then $\#\{(-,-)\} > 0$; that is, there exists points of type $\{(-,-)\}$* (which does not happen for semigroups with two generators).

Example 3. $N(5,17,19) = 34$. The numbers of type $(+)$ are: $\{0,5, 10,15,17,19,20,22,24,25,27,29,30,32\}$. There are 14 of them.

The numbers of type $(-)$ are: $\{1,2,3,4,6,7,8,9,11,12,13,14,16, 18,21,23,26,28,31,33\}$. There are 20 of these.

The numbers of type $(-,-)$ are: $\{2,7,12,21,26,31\}$. There are 6 of these.

The 6 numbers of type $(-,-)$ form a remarkable diagram:

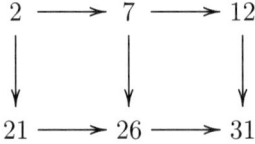

where "$\longrightarrow$" means "add $a = 5$", and "$\downarrow$" means "add $c = 19$".

Let m be the smallest number of type $(-,-)$, and let M be the largest number of this type.

Conjecture 1. *The difference between M and m is always a number of type $(+)$.*

Example. In the previous example $M = 31$, $m = 2$, $M - m = 29 = 2a + c$ (because the rectangular diagram above has length 2 in the a-direction and height 1 in the c-direction).

The conjecture is based on the fact that in numerous examples the numbers of type $(-,-)$ form rectilinear, rectangular, or parallelepipedal diagrams similar to the rectangle of $\{(-,-)\}$-points shown above.

Example. $N(10,13,48) = 56$, $\{(-,-)\} = \{8 \to 21 \to 34 \to 47\}$, a segment in the direction of $(b = 13)$ with four vertices.

Example. $N(9,13,19) = 44$, $\{(-,-)\} =$
$$
\left\{
\begin{array}{c}
\begin{array}{ccc}
20 & \longrightarrow & 29 \\
\nearrow & & \nearrow \\
1 \longrightarrow 10 & & \downarrow \\
\downarrow \quad\; \downarrow & & \\
& 33 \longrightarrow 42 & \\
\downarrow \nearrow & \downarrow \nearrow & \\
14 & \longrightarrow & 23
\end{array}
\end{array}
\right\}.
$$

This is a parallelepiped with 8 vertices and with edges "$\longrightarrow$" in the a-direction, "$\downarrow$" in the b-direction, and "$\nearrow$" in the c-direction, all of length 1.

Conjecture 2. *The set of points of type $\{(-,-)\}$ always forms a parallelepiped (of dimensions 1, 2, and 3), whose edges connect vertices m and M with directed segments showing an increase of vertices by a, by b, or by c. Namely, if $M = m + (ua + vb + wc)$, then the edges in the a-direction have length u, the edges in the b-direction have length v, and the edges in the c-direction have length w.*

I formulated this conjecture not because I can prove it, but because I hope that experiments done by the Dubna school (perhaps with computers) will help either to find a counter-example, or to supplement the list I have of confirmed examples which render the conjecture more convincing.

The hopes of using the proposed structure of sets $\{(-,-)\}$ described above for bounding the number of $(+)$-points from below are founded on the hopes of deriving an upper bound for the number of $(-,-)$-points from this structure.

For example, *in order to prove the bound $\#\{(+)\} > \epsilon N$ it would suffice to show that $\#\{(-,-)\} \le (1 - 2\epsilon)N$.*

In examples known to me the inequality $\#\{(-,-)\} \leq N/3$ is always satisfied (and even a bit more) so that the inequality $\#\{(+)\} \leq N/3$ (and even a stronger inequality) is also satisfied. But in the general case such bounds have not been proven.

Several analogues to these hypothetical bounds are given by the following

Problem. *What part of the volume of the tetrahedron can be completely covered by a parallelepiped contained within it?*

Conjecture 3. $\dfrac{(\textit{The volume of the parallelepiped})}{(\textit{The volume of the simplex in } \mathbb{R}^n)} \leq \dfrac{n!}{n^n}.$

Example. For $n = 2$ the right hand side is equal $1/2$, and a parallelogram contained in the triangle can cover half its area. It is easy to prove that it cannot cover more than half its area.

For $n = 3$ the right side $n!/n^n$ is equal to $2/9$. It is not hard to cover this much of the volume of a pyramid with a parallelepiped contained within it. For example, for the pyramid $\{x \geq 0, \; y \geq 0, \; x + y + z \leq 1\}$, two ninths of its volume are covered by the cube $\{0 \leq x \leq 1/3, \; 0 \leq y \leq 1/3, \; 0 \leq z \leq 1/3\}$. A proof that it is impossible to cover more than $2/9$ of the volume, is not so simple (but, I would think, not so difficult for the students at Dubna).

The hope of using this problem to find an upper bound for the number of integer points in the required parallelepiped $\{(-,-)\}$ is based on the fact that results in the convex geometry of polyhedra usually have integer analogues in the geometry of Minkowski (where the role of the volume of the polyhedron is played by the number of its integer points).

Remark. The Frobenius number N is not the number of integer points in the pyramid $\{x \geq 0, \; y \geq 0, \; z \geq 0, \; ax + by + cz \leq N - 1\}$ into which we want to place the parallelepiped $\{-,-\}$. Rather, it is a peculiar analogue of the altitude of this pyramid (dropped to the "hypotenuse" from vertex O). This same "altitude" is also an upper bound for the number of points of type $(-,-)$.

Therefore for the proof of upper bounds for the number of points of type $(-,-)$ using the Frobenius number $N(a,b,c)$ Conjecture 3 is not sufficient (although it may be a necessary step in this proof).

But the volume of the pyramid is still bounded from above by a quantity of order N (as was seen in the proof of the lower bound $N \geq \text{const.} \cdot \sigma^{3/2}$) for the Frobenius number given in Section 4.3). The altitude $\#\{(-,-)\}$ of the conjectural parallelepiped is not greater than the number of its integer points, which brings to mind its volume, which is bounded by Conjecture 3. Therefore Conjecture 3 gives us hope for an upper bound for the number of points of type $(-,-)$ by a certain fraction of the Frobenius number,

$$\#\{(-,-)\} \leq (1 - 2\varepsilon)N.$$

It would be interesting to study not just the total mass of points of a semigroup on the segment up to the Frobenius number, but also to characterize the distribution of the points of the semigroup on this segment.

Empirically, the density of this distribution often turns out to grow more or less like some power (with the expected exponent 2, which becomes $n - 1$ for a semigroup with n generators).

The exponent $n - 1$ can be explained by the relation $d\ell^n = n\ell^{n-1}d\ell$ (Figure 4 in Section 4.). But resonances (like the relationship $P(a,b,c) = P(a,b)$ for $c - a \mid b$) break the connection between the number of points in the projection of the integer points of an n-dimensional pyramid and ℓ^n. Therefore it would not be easy to turn the observation mentioned of the growth in density into a theorem.

I formulated the program above for lower bounds on points of the type $(+)$ for the case of a semigroup with three generators. But an analogous program is also suitable for $n > 3$ generators (with the conjectured bound $\#\{(+)\} \geq N/n$ and with the conjecture about the growth of the density which I have proposed in my work of 1999, cited above).

Editor's notes

1. There are more recent results concerning this conjecture. I would mention the article by A. V. Ustinov [19*]. The author proves that the average of the numbers $N(a, b, c)$ is asymptotically close to $(8/\pi)\sqrt{abc}$.

2. Explicit formulas for computing Frobenius numbers have been known for a long time. The best known result is contained in an article by O. J. Rødseth, [17*]. The author proves a formula for the numbers $N(a, b, c)$ valid in the case when every pair of the positive integers a, b, c (not just all three of them) have no common divisors greater than 1. It is interesting that Rødseth's formula also uses some non-traditional continued fractions. (Notice that there exist generalizations of this formula to the case when only all three integers a, b, c have no common divisors, and to the case of N with more than three arguments.)

   Let us describe Rødseth's formula. Let (a, b, c) be a triple of pairwise relatively prime positive integers. We will assume that $a > b > c$. An elementary theorem from number theory states that there exists a unique integer ℓ such that $1 < \ell < a$ and $b\ell \equiv c \bmod a$. It is also true that there exists a unique development of a/ℓ into a "continued fraction"

$$\frac{a}{\ell} = a_1 - \cfrac{1}{a_2 - \cfrac{1}{a_3 - \cfrac{\ddots}{ - \cfrac{1}{a_m}}}}$$

where $a_1 \geq 2, \ldots, a_m \geq 2$. Consider the numbers $q_0, \ldots, q_{m+1}$ and $s_0, \ldots, s_{m+1}$ defined by the recursive formulas

$$q_0 = 0, \quad q_1 = 1, \quad q_j = q_{j-1}a_{j-1} - q_{j-2} \text{ for } 2 \leq j \leq m+1$$
$$s_{m+1} = 0 \quad s_m = 1, \quad s_j = s_{j+1}a_{j+1} - s_{j+2} \text{ for } m-1 \geq j \geq 0.$$

It is not hard to prove that $q_{m+1} = s_0 = a$, $s_1 = \ell$ and

$$\infty = \frac{s_0}{q_0} > \frac{s_1}{q_1} > \cdots > \frac{s_m}{q_m} > \frac{s_{m+1}}{q_{m+1}} = 0.$$

Hence, there exists a unique v such that

$$\frac{s_v}{q_v} > \frac{c}{b} \geq \frac{s_{v+1}}{q_{v+1}}.$$

The formula says that

$$N(a, b, c) = f(a, b, c) - (a + b + c) + 1$$

where

$$f(a, b, c) = bs_v + cq_{v+1} - \min(bs_{v+1}, cq_v).$$

Example 1. Let $a = 11, b = 5, c = 3$. Then $\ell = 5$,

$$\frac{11}{5} = 3 - \cfrac{1}{2 - \cfrac{1}{2 - \cfrac{1}{2 - \cfrac{1}{2}}}},$$

$$q_0 = 0, \quad q_1 = 1, \quad q_2 = 3, \quad q_3 = 5, \quad q_4 = 7, \quad q_5 = 9, \quad q_6 = 11,$$
$$s_0 = 11, \quad s_1 = 5, \quad s_2 = 4, \quad s_3 = 3, \quad s_4 = 2, \quad s_5 = 1, \quad s_6 = 0,$$

and $v = 2$:

$$\frac{s_2}{q_2} = \frac{4}{3} > \frac{3}{5} \geq \frac{3}{5} = \frac{s_3}{q_3}.$$

The formula says that

$$f(11, 5, 3) = 5 \cdot 4 + 3 \cdot 5 - \min(5 \cdot 3, 3 \cdot 3) = 20 + 15 - 9 = 26$$

and

$$N(11, 5, 3) = 26 - (11 + 5 + 3) + 1 = 8$$

which agrees with Arnold's computations (see Section 4.1).

Example 2. Let $a = 19, b = 15, c = 7$. Then $\ell = 3$,

$$\frac{19}{3} = 7 - \cfrac{1}{2 - \cfrac{1}{2}},$$

$$q_0 = 0, \quad q_1 = 1, \quad q_2 = 7, \quad q_3 = 13, \quad q_4 = 19,$$
$$s_0 = 19, \quad s_1 = 3, \quad s_2 = 2, \quad s_3 = 1, \quad s_4 = 0,$$

and $v = 1$:

$$\frac{s_1}{q_1} = \frac{3}{1} > \frac{7}{15} \geq \frac{2}{7} = \frac{s_2}{q_2}.$$

The formula says that

$$f(19, 15, 7) = 15 \cdot 3 + 7 \cdot 7 - \min(15 \cdot 2, 7 \cdot 1) = 45 + 49 - 7 = 87$$

and

$$N(19, 15, 7) = 87 - (19 + 15 + 7) + 1 = 47$$

which also agrees with Arnold's computations (see Section 4.5).

Example 3. Let $a = 29, b = 7, c = 5$. Then $\ell = 9$,

$$\frac{29}{9} = 4 - \cfrac{1}{2 - \cfrac{1}{2 - \cfrac{1}{2 - \cfrac{1}{3}}}},$$

$$q_0 = 0, \quad q_1 = 1, \quad q_2 = 4, \quad q_3 = 7, \quad q_4 = 10, \quad q_5 = 13, \quad q_6 = 29,$$
$$s_0 = 29, \quad s_1 = 9, \quad s_2 = 7, \quad s_3 = 5, \quad s_4 = 3, \quad s_5 = 1, \quad s_6 = 0,$$

and $v = 2$:

$$\frac{s_2}{q_2} = \frac{7}{4} > \frac{5}{7} \geq \frac{5}{7} = \frac{s_2}{q_2}.$$

The formula says that

$$f(29, 7, 5) = 7 \cdot 7 + 5 \cdot 7 - \min(7 \cdot 5, 5 \cdot 4) = 49 + 35 - 20 = 64$$

and

$$N(19, 15, 7) = 64 - (29 + 7 + 5) + 1 = 24$$

which again agrees with Arnold's computations (see Section 4.5).

The relations between Rødseth's formula and Arnold's geometric approach have been studied by A. V. Ustinov, who was able to deduce Rødseth formula from Arnold-like geometric constructions. See A. V. Ustinov, [**20***].

Bibliography

Note: references marked with an asterisk were added by the editor.

[1]   V. I. Arnold. *A Mathematical Trivium*, Russian Math. Surveys **46** (1991), 271–278.

[2]   V. I. Arnold. *Weak Asymptotics of the Numbers of Solutions of Diophantine Problems*, Funct. Anal. Appl. **33** (1999), no. 4, 292–293

[3]   V. I. Arnold. *Geometry and dynamics of Galois fields.* Russian Math. Surveys **59** (2004), no. 6, 1029–1046.

[4]   V. I. Arnold. *Statistics of Young Diagrams of Cycles of Dynamical Systems for Finite Tori Automorphisms*, Moscow Math. J. **6** (2006), no. 1, 43–46.

[5]   V. I. Arnold. *Geometry and the Growth Rate of Frobenius Numbers of Additive Semigroups*, Math. Phys. Anal. Geom. **9** (2006), 95–108.

[6]   V. I. Arnold. *Statistics and the Classification of the Topologies of Periodic Functions and Trigonometric Polynomials*, Proc. Steklov Inst. Math (2006), Dynamical Systems: Modeling, Optimization, and Control, suppl. 1, S13–S23.

[7]   V. I. Arnold. *Complexity of finite sequences of zeros and ones and geometry of finite spaces of functions*, Funct. Anal. Other Math. **1** (2006), no. 1, 1–18.

[8]   V. I. Arnold. *Smooth functions statistics*, Funct. Anal. Other Math. **1** (2006), no. 2, 111–118.

[9]   V. I. Arnold. *Dynamics, statistics and projective geometry of Galois fields.* MCNMO, Moscow. 2006; English translation: Cambridge University Press, Cambridge, 2011.

[10]   V. I. Arnold. *Arithmetical Turbulence of Self-similar Fluctuations: Statistics of Large Frobenius Numbers of Additive Semigroups of Integers*, Moscow Math. J. **7** (2007), 173–193.

[11]   V. I. Arnold. *Topological Classification of Trigonometric Polynomials Related to the Affine Coxeter Group $\tilde{A}_2$*, Proc. Steklov Inst. Math. **258** (2007), no. 1, 3–12.

[12*] D. Fuchs and S. Tabachnikov. *Mathematical Omnibus*. Providence RI, AMS, 2007.

[13*] A. Garber. *Graphs of difference operators of p-ary sequences*, Funct. Anal. Other Math. **1** (2006), no. 2, 159–174.

[14*] O. Karpenkov. *On examples of difference operators for $\{0,1\}$-valued functions on finite sets*, Funct. Anal. Other Math. **1** (2006), no. 2, 175–180.

[15*] V. Kharlamov and O. Viro. *Easy reading on topology of plane real algebraic curves*, preprint.

[16]   L. Nicolaescu. *Morse Function Statistics*, Funct. Anal. Other Math. **1** (2006), no. 1, 97–103.

[17*] O. J. Rødseth. *On a linear Diophantine problem of Frobenius*, J. Reine und Angew. Math. **301** 1978, 171–178.

[18*] K. Rosen. *Elementary number theory*. Addison-Wesley.

[19*] A. V. Ustinov. *Solution of Arnold's problem on weak asymptotics for Frobenius numbers with three arguments*, Sbornik Math. **200** (2009), 597–627.

[20*] A. V. Ustinov. *Geometric proof of Rødseth's formula for Frobenius numbers*, Proc. Steklov Inst. Math., **276** (2012), 275–282.

Page ii constitutes the beginning of this copyright page.

Library of Congress Cataloging-in-Publication Data

Arnol′d. V. I. (Vladimir Igorevich), 1937–2010.
    [Eksperimentalnoe nablyudenie matematicheskikh faktov. English]
    Experimental mathematics / V.I. Arnold ; translated by Dmitry Fuchs and Mark Saul.
        pages cm. — (MSRI mathematical circles library ; 16)
    Includes bibliographical references.
    ISBN 978-0-8218-9416-3 (alk. paper)
    1. Experimental mathematics. 2. Geometry, Algebraic. 3. Combinatorial analysis. 4. Functions. I. Fuchs, D. B. II. Saul, Mark E. III. Title.

QA9.A7613 2015
510.72′4—dc23
                                                                            2015009318